ROBOTICS
APPLICATIONS AND SOCIAL
IMPLICATIONS

ROBOTICS
APPLICATIONS AND SOCIAL IMPLICATIONS

Robert U. Ayres
Steven M. Miller

with contributions from
Vary Coates
David Criswell
Leonard Lecht
John Molburg
Paul Wright

Ballinger Publishing Company
Cambridge, Massachusetts
A Subsidiary of Harper & Row, Publishers, Inc.

338.47629892
A985

International Standard Book Number: 0-88410-891-0

Library of Congress Catalog Number: 82-13881

Printed in the United States of America

Library of Congress Cataloging in Publication Data

Ayres, Robert U.
    Robotics, applications and social implications.

    Includes bibliographies and index.
    1. Robot industry. I. Miller, Steven M. II. Title.
HD9696.R622A97   1982   338.4'7629892   82-13881
ISBN 0-88410-891-0

CONTENTS

List of Figures   vii
List of Tables   xi
Acknowledgments   xvii

1. **Background**   1

    Scope of the Problem   1
    Recent History and Forecasts   4
    The Concept of Technology Assessment   8
    References   12

2. **Robot/CAM Technology**   15

    Robotics and the Industrial Revolution   15
    Technical Limits of Robots   22
    Robot Technology: A Review   29
    Applications in Standard Industrial Tasks   41
    Integration of Robots into CAD/CAM Systems in
    Metalworking   48
    Applications in Other Manufacturing Sectors 57

Future Technological Development   60
References   63

3. **Costs and Benefits of Robots in Manufacturing**   69

Motivations for Retrofitting Robots in Existing
Factories   70
Decision Criteria To Implement Robots   78
Flexible Manufacturing Technologies   84
Organizational Adaptation   93
Appendix   98
References   123

4. **Applications of Robots in Hazardous or Inaccessible
    Environments**   127

Robots in Underground Coal Mining   130
The Space Environment   151
Robots in the Ocean   175
Other Future Uses of Robots   182
References   183

5. **Employment Impact of Robotization**   187

Aggregate Trends in Manufacturing Employment   188
Occupational Employment in Manufacturing   191
Estimating Potential Displacement   197
Projected Employment Growth in Manufacturing,
1980–90   205
Regional Implications   212
Union Response to Technological Change   217
Job Openings   226
References   238

6. **Productivity Impacts**   241

    Overview of Recent Productivity Trends   242
    Labor Productivity and Unit Labor Cost   245
    Technical Knowledge, Productivity, and Capital   249
    Low-, Mid- and High-Volume Production   259
    Cost versus Batch Size   262
    Narrowing the Gap Between Batch and Mass
    Production   266
    Potential Impacts on Labor Cost   269
    Potential Impacts on Capacity   272
    Potential Impacts on Capital Costs   283
    Consequences of Dramatic Improvements in
    Productive Capabilities   286
    Concluding Comments   297
    References   299

7. **Policy Implications**   303

    Human Resources Policy   306
    National R&D Policy   312
    Corporate R&D Policy   313
    Corporate Compensation and Promotion Policy   315
    References   318

8. **The Ultimate Robot**   319

    References   333
    Index   335

LIST OF FIGURES

Figure 1–1.   Estimates of the U.S. Robot Population: 5
              1970–81.

Figure 1–2.   Impact of Technology Innovation on Us-   10
              er Innovations.

Figure 2–1.   Architecture and Work Volume: Carte-   31
              sian, Cylindrical, and Spherical.

Figure 2–2.   Architecture and Work Volume: 32
              Revolute.

Figure 2–3.   Use of Via Points to Speed Up Manipula-   37
              tor Motions.

Figure 2–4.   Product Flow in a Manufacturing Cell.   49

Figure 2–5.   Part Drilled and Routed by Robot.   51

Figure 2–6.   Technological "Core" of a Robot-Inte-   53
              grated Manufacturing Cell.

Figure 2–7.   Flexible Computerized Manufacturing   55
              System.

Figure 2–8.   Schematic View of Computer-Robot Inte-   58
               grated Manufacturing Cell for Powder
               Metallurgy.

Figure 3–1.   A Comparison of Robot and Human   72
               Costs.

Figure 3–2.   Discounted Present Worth of Costs and   82
               Benefits of a Robot.

Figure 3–3.   Efficiency Versus Flexibility Tradeoffs.   89

Figure 4–1.   Technology Choice.   128

Figure 4–2.   U.S. Coal Production by Method.   131

Figure 4–3.   Coal Use Projections, Various Sources.   133

Figure 4–4.   Energy Use Projections by Primary Fuels   134

Figure 4–5.   Productivity of Coal Mines.   135

Figure 4–6.   Coal Prices.   136

Figure 4–7.   Coal Mine Fatality Rates.   137

Figure 4–8.   Mechcnically Anchored and Resin-   143
               Anchored Roof Bolts.

Figure 4–9.   The Joy Pushbutton Miner.   146

Figure 4–10.   Spacecraft Automation Trends.   153

Figure 4–11.   Trend of Decision Allocation between   154
               Humans and Spacecraft.

Figure 4–12.   The Growth of the INTELSAT.   157

Figure 4–13.   Satellite Availability On-Orbit.   161

Figure 4–14.   Pivoting Arm On-Orbit Servicer.   162

Figure 4–15.   Assembly/Disassembly Station (A/DS)   168
               in the Bay of the Space Shuttle.

Figure 4–16.   Automated Impact Molder Using Robots   173
               for Trimming and Inspection of Parts.

Figure 4–17.   Artist's Conception of a Construction   174
               Unit for Space (CUS).

Figure 5–1.   Distribution of the Manufacturing   195
              Workforce by Sex and Race, 1980.

Figure 6–1.   Total Factor Productivity in the U.S. 246
              Economy, by Sector.

Figure 6–2.   Trade Trends for Light and Heavy Ma- 257
              chinery and Its Manufactures.

Figure 6–3.   Trade Trends for Other Dominant Cate- 257
              gories of Manufacturers.

Figure 6–4.   Average Cost Versus Batch Size.   263

Figure 6–5.   Average Cost Proportions in Metalwork- 271
              ing (SIC 33–38), 1977.

Figure 6–6.   Average Cost Proportions in All Manu- 271
              facturing (SIC 20–39), 1977.

Figure 6–7.   Breakdown of Theoretical Capacity in   274
              Low-, Mid-, and High-Volume Manufac-
              turing.

Figure 6–8.   Potential Impact of Robotics and CAM   287
              on Cost of Batch Processes.

Figure 7–1.   Effects of Interruptions on a Learning 317
              Curve.

Figure 8–1.   Proposed System for Lunar Extraction of 327
              Oxygen and Metals.

Figure 8–2.   Stationary Universal Construction Unit 329
              (SUC).

Figure 8–3.   Mobile Universal Construction Unit   330
              (MUC).

LIST OF TABLES

Table 1–1.   International Robot Population: February, 1982.   7

Table 2–1.   Tasks that Robots Can Do and May Be Able to Do.   25

Table 2–2.   Distribution of U.S. Robot Population by Application.   42

Table 2–3.   Comparison of Manual Manufacturing Steps Elimination by Various Degrees of Computer Control.   54

Table 4–1.   Percentage of Underground Workforce by Occupation.   149

Table 4–2.   Communication Satellites of the World.   156

Table 4–3.   Typical Failures of Communications Satellites.   160

Table 4–4.   Manufacturing Processes Applicable to Space Based on Terrestrial Experience.   170

Table 4–5.   Scenario for the Growth of Space Manufacturing and the Use of Robotics   171

Table 4–6.   Remote-Operated Vehicle Work Cat-   179
               egories.

Table 5–1.   Number of Persons on Payrolls of Nonag-   189
               ricultural Establishments, 1948–80.

Table 5–2.   Percentage of Workforce Employed in   190
               Goods and Service Sectors.

Table 5–3.   Nonproduction Workers as Percentage of   191
               Total Employment by Major Sector.

Table 5–4.   Employment of Production Workers,   192
               1980: Metalworking and Total Manufac-
               turing.

Table 5–5.   Representative Occupational Titles for   193
               Craft Workers, Operatives, and Laborers.

Table 5–6.   Analysis of the Potential for Robotic   200
               Substitution by Task.

Table 5–7.   Skill Classification Used in the *Diction-*   202
               ary of Occupational Titles.

Table 5–8.   Breakdown of the Work Requirements   203
               for Inspectors.

Table 5–9.   Potential for Robotization.   206

Table 5–10.   Employment in Manufacturing and   209
               Metalworking Industries, 1980 and Pro-
               jected 1990 (Bureau of Labor Statistics
               Scenarios.)

Table 5–11.   Projected Employment Growth Versus   211
               Potential Robotic Replacement.

Table 5–12.   Value Added and Employment Growth in   213
               Manufacturing by Geographic Division.

Table 5–13.   Distribution of Employment in Durable   214
               Goods Industries, by Major Region,
               1968–78.

Table 5–14.   Employment of Production Workers in   215
               Metalworking: Leading States, 1977.

Table 5–15. Persons Moving to and from Each Major   217
Region, 1973–76.

Table 5–16. Wage and Salary Workers Represented by   218
Labor Organizations, May 1980.

Table 5–17. Major Unions Representing Workers in   219
the Metalworking Industries.

Table 5–18. Age Distribution of the Manufacturing   221
Workforce, 1980.

Table 5–19. Collective Bargaining Provisions Rele-   223
vant to Technological Change.

Table 5–20. Movements out of Occupational Groups.   227

Table 5–21. Projected Employment Changes for   228
High-Growth Industries, 1979–90.

Table 5–22. Projected Employment Changes for   230
High-Growth Occupations, 1978–90.

Table 5–23. Definitions of Skill Levels for a Job's Re- 233
lationship to *Things*.

Table 5–24. Examples of Range of Physical Skill   235
Levels for Operatives.

Table 5–25. Supplementary Skill Classifications for   237
the DOT.

Table 6–1.   Estimates of Total Factor Productivity   244
Growth Rates.

Table 6–2.   Output, Compensation, and Unit Labor   247
Costs in Manufacturing, 1950–79.

Table 6–3.   Import Measures for Selected Metal-   248
working Products.

Table 6–4.   Output Per Hour, Hourly Compensation,   250
and Unit Labor Costs in Manufacturing,
1960–80, Average Annual Rates of
Change.

Table 6–5.   Contributions to the Slowdown in the   254
              Growth of Labor Productivity in the Pri-
              vate Business Sector, 1965–78.

Table 6–6.   Age Distribution of Machine Tools in   256
              Use.

Table 6–7.   Overview of Discrete Parts Production.   259

Table 6–8.   Product-Process Overview of Piece,   260
              Batch, and Mass Production.

Table 6–9.   Flexibility of Piece, Batch, and Mass Pro-   261
              duction Systems.

Table 6–10.   Unit Cost Index of Selected Steel Prod-   265
              ucts.

Table 6–11.   Potential Cost Savings by Eliminating   268
              the Batch-Mass Gap.

Table 6–12.   Estimating the Impacts of Robotics   268
              Technology on the Components of Unit
              Cost.

Table 6–13.   Estimates of Average Machine Tool Utili-   273
              zation in the Metalworking Industries,
              1977.

Table 6–14.   Estimates of Planned Production Time in   275
              Low-, Mid-, and High-Volume Metal
              Fabricating Manufacturing.

Table 6–15.   Potential Percentage Increases in Output   277
              from Utilizing Lost Time.

Table 6–16.   Potential Percentage Output Increases   279
              from Recouping Non-Productive Time
              during Planned Operations: High-Volume
              Manufacturing.

Table 6–17.   Potential Output Increases from Recoup-   280
              ing Non-Productive Time during Planned
              Operations: Mid-Volume Manufacturing.

Table 6–18.   Potential Output Increases from Recoup-   281
              ing Non-Productive Time during Planned
              Operations: Low-Volume Manufacturing.

Table 6–19.   Summary of Potential Impacts on Capac-   282
              ity.

Table 6–20.   Inventory as a Fraction of Value of Ship-   284
              ments.

Table 6–21.   Fraction of Metalworking Shipments to   290
              Intermediate Uses and Capital Invest-
              ment.

Table 7–1.   CETA-Sponsored Training and Retrain-   308
              ing Programs.

Table 7–2.   Enrollments and Completions in Public   309
              Vocational Education in Selected Metal-
              working Occupations: National Totals,
              Fiscal Year 1978.

ACKNOWLEDGEMENTS

While the bulk of the material in this book was finally written by the two principal authors, we acknowledge significant contributions from several others. Vary Coates, a consulting political scientist and adjunct professor at Carnegie-Mellon University (CMU), wrote a draft chapter that was later broken up and used in several places in the book. Paul Wright, Associate Professor of Mechanical Engineering at CMU and a key member of the University's Robotics Institute, contributed several draft sections of Chapter 2. Professor Paul Goodman and Professor Linda Argote, of the Graduate School of Industrial Administration at CMU, revised and extended our draft section on organizational adaptations in Chapter 3, based on their recent field study of how workers within a plant reacted to robot use. John Molburg, a consulting mechanical engineer, did much of the research and an initial draft on applications of robots to coal mining in Chapter 4. David Criswell, of the California Space Institute, University of California, Le Jolla, provided us with an overview of robot applications in the oceans and in space for this chapter. He is also a strong advocate of the self-reproducing robot factory concept, especially as it might be applicable to manufacturing on the lunar surface. Leonard Lecht, a consulting labor economist (formerly with the Conference Board) wrote the first draft of Chapter 5. Wilbur Steger, of CONSAD Research Corporation, Pittsburgh, helped us to draft chapter 6. In addition, we borrowed ideas along the way from Steve Goldstein of MITRE Corporation, Leonard Lynn of the CMU Social Science Department, Dr. Hans Moravec of the Robotics Institute, Professor Allen Newell, Professor Raj Reddy (Director of the Robotics Institute), and Professor Herbert Simon, all of the CMU Computer Science Department

We also freely borrowed data from a "project report" prepared in spring semester, 1981, by a group of CMU graduate students in the School of Urban and Public Affairs and undergraduates in the Department of Social Science and the Department of Engineering and Public Policy. This report, entitled "The Impact of Robotics on the Workforce and Workplace," was managed and edited by Steve Miller. The senior author of this book (Robert U. Ayres) taught a course on the social impacts of robotics in the fall term of 1981 to a group of business students from the Graduate School of Industrial Administration at CMU. He learned much from these students.

The manuscript was extensively reviewed and revised at several stages of its evolution. A large number of people supplied us with information on the history of robotics. Richard Hohn and Richard Eby, from Cincinnati Milicron, Tali Cepuritis, a patent attorney for Prab Robots Inc., and George Devol, one of the inventors of the industrial robot, were especially helpful. Many members of the CMU Robotics Institute (including Gerald Agin, Hans Moravec, and Marc Raibert) and Richard Hohn, from Cincinnati Milacron, reviewed parts of the sections on robot technology. Edwald Heer, from the Jet Propulsion Lab, was most helpful in reviewing the chapter on robotic applications in hazardous or inaccessible environments. William Gervarter, formally with NASA, and now with the National Bureau of Standards, and James Romero from NASA headquarters, also provided some comments on the section on space applications. Chuck Thorpe, of the CMU Robotics Institute, reviewed the section on undersea applications.

Many divisions of the Bureau of Labor Statistics (BLS), U.S. Department of Labor, supplied us with the labor force information used in Chapter 5. We are grateful to Neil Rosenthal, assistant commissioner for employment projections for supplying us with unpublished information, and for having the employment chapter read over by his staff. Al Lorente, of the Skilled Trades Division of the United Auto Workers, and William Bittle, of the International Association of Machinists, frequently provided us with information on union policies. Randy Norsworthy, former chief of productivity research at the Bureau of Labor Statistics and now chief of economic studies at the Bureau of the Census, was especially helpful in reviewing our general discussion of productivity in Chapter 6. John Kendrick, professor of Economics at George Washington University, also provided some insightful comments. Charles Carter, Technical Director for Cincinnati Milacron Inc., and David Lee, senior manufacturing engineer with the Ford Motor Company, critiqued our discussion of the potential impacts of robotics on productivity. Denis Ceechini and others at the General Motors Manufacturing Development Center also provided us with some useful insights regarding robotics, productivity, and economic growth.

We wish to thank Professor Allen Newell and Professor Herbert Simon for reviewing Chapter 8, The Ultimate Robot, and for their respective views on the future of robotics.

We gratefully acknowledge financial support from the CMU Robotics Institute, especially during 1980-81. Professor Raj Reddy commissioned a technology assessment in the spring of 1980 which eventually evolved into this book. In addition, the Institute has made available computer time for text processing, and assisted in producing much of the graphic work for the book. They have also helped us keep to abreast of developments in robotics in many ways. Todd Simonds, assistant director and head of the Industrial Affiliates Program, has been most helpful in coordinating our research with the Institute's activities.

Special thanks are due to the Westinghouse Electric Corporation, not only for its original financial support of the CMU Robotics Institute, but also for its continuing cooperation with us personally in terms of providing data and advice from a number of its executives and technical experts.

The Robot Institute of America, in the person of its executive staff, Bernie Sallott and Lori Mei, have also been very helpful at several stages, as have many of its corporate members.

The Department of Engineering and Public Policy at CMU has supplied much of the administrative support required to produce the manuscript, and has supported us financially during second year of our work. Our department head, M. Granger Morgan, has continuously encouraged this project and all our background research. Sandy Rocco, the department's administrative head, smoothed the way. Cathy Hill typed the first several drafts of the manuscript, and Gema Barkanic produced the final draft on the *Scribe* word processing system used at Carnegie-Mellon University. Unilogic, Inc., of Pittsburgh, owners of *Scribe*, provided us with much assistance in producing the manuscript.

Opinions expressed and conclusions reached in this book are our own, and do not necessarily reflect the views of the institutions which provided financial support, or of the many people who assisted us.

Robert U. Ayres and Steven M. Miller
Pittsburg, June, 1982

1 BACKGROUND

SCOPE OF THE PROBLEM

The term "robot" allegedly stems from a Czech word "robotnik," meaning worker or serf. It was first introduced into the popular language by the playwright Karel Capek, in his 1920 play *R.U.R.* (Capek 1973). This work of fiction imaginatively explored some speculative social implications of the development of synthetic (android) workers, including possible obsolescence and replacement of the human race! The robot theme was explored in a different fashion by Isaac Asimov in science fiction of the early 1950s. Since then there has been an efflorescence of literature dealing with the potential implications of intelligent and—ultimately—"conscious" computers, and their relations with mankind. One of the best known examples is the malign (insane?) computer HAL in the movie *2001*; one of the most convincing and technically plausible is the novel *The Genesis Machine* by James P. Hogan (1979).

Popular books on the subject, such as Malone (1978) and Reichardt (1978), tend to define robots as "anthropomorphic automata." This opens the field to a vast range of actual or literary artifacts designed for entertainment or mystification, going back hundreds and even thousands of years. Robots to do work appeared only recently in literature, however, and even more recently in practice. Working robots, in fact, owe little to literary imaginings—most of which were wildly unrealistic.

Clearly we must impose strict limits on our analysis. The "ultimate conflict" issue is one that we can safely ignore for the time being. Far from being electromechanical (or biomechanical) avatars of humans, industrial robots today—and for the foreseeable future—are simply mechanical transfer and manipulator devices with some degree of generalized (multitask) capability and programmability. Sensory feedback is becoming available, and elements of artificial intelligence will eventually be added. To put the situation in perspective: robots can already perform some very simple manipulative tasks as fast as or faster, and can repeat them more accurately, than humans. Robots can also function in some environments antithetical to humans. In a decade they will presumably be capable of handling many additional tasks of the kind that human hands (with tools) normally perform. They will accordingly play a significant role in industry. Nevertheless, even the enormous capabilities of present-day electronic microprocessors, linked with artificial senses and mechanical manipulators, provide capabilities that are competitive with human skills only in the context of a few specialized repetitive industrial tasks. (Table 2-1 suggests the sorts of tasks that robots now, and in the future, may be capable of doing.)

Dramatic improvements in all three of these areas—mechanical capabilities, sensory feedback controls, and artificial intelligence—will be required before robot manufacturers can aspire to replace a significant fraction of the present industrial labor force (e.g., assemblers). Moreover, this is only possible because many factory jobs—particularly those classed as "operatives"—involve trivial tasks in an artificially simplified environment. It is not unreasonable to compare the motor and sensory capabilities of an industrial robot (disregarding strength) with those of a grasshopper or perhaps a turtle. To put it another way, robots do not fully utilize the highly developed sensory-motor coordination and general purpose rapid decisionmaking skills that are required for survival in a diverse, variable and unpredictable environment. Viewed in this light, any threat by robots to the human race, as such, must be regarded as very remote.

Social impacts of a less dramatic sort—both beneficial and harmful—are obviously not remote, however. On the positive side, robots (and computers) may increase both labor and capital productivity in the manufacturing sectors of the economy. The traditional distinction between batch and mass production is likely to become blurred. This would have significant benefits to the economy, especially capital goods producers and consumers. The indirect impact on the organization of production and on technological innovation could be even more important in the long run.

The frequency of certain kinds of industrial and mining accidents will very likely be reduced sharply. Some other occupational hazards, such as exposure to radioactivity, toxic chemicals, cotton lint, asbestos fiber, or coal dust may also be cut significantly. Moreover, robots will enable us to effectively explore and exploit environments that are inaccessible or hostile to humans. Examples include very deep surface mines, contaminated areas, oceans, earth orbit, the lunar surface, asteroids, and other planets of the solar system. This new capability will greatly extend the range of possible search for resources, among other important consequences. Finally, robots may eventually be capable of undertaking various household tasks and becoming, in effect, a super consumer appliance.

On the negative side, robots can be expected to eliminate some existing jobs, especially in the semiskilled categories. Almost certainly this will mean displacement, and in some cases unemployment, for many workers. The severity of the resulting social problems depends, of course, on social policy. Given a sufficiently enlightened, humane, and far-sighted policy, the problem might be mitigated by provision of transitional counseling, retraining, and perhaps relocation assistance. On the other hand, if the widespread introduction of robots (and other forms of computer assistance) in manufacturing occurs without any advanced planning, the results could be a severe deterioration of industrial labor relations, increased union militancy, and a rash of ill-advised legislative "remedies" restricting economic freedom at a time when flexibility may be the key to survival. Even worse, some social critics envision the possibility of a new antitechnological backlash accompanied by social pathologies from anomie to organized industrial sabotage.

RECENT HISTORY AND FORECASTS

Industrial robots first appeared on the market in 1959 and came into use in the United States very slowly during the 1960s. There were probably only a few hundred in use by 1970. Throughout the 1970s, there were yearly predictions by robotics manufacturers and market analysts that the technology was about to "take off." This did not actually happen in the United States until 1979 (although other countries moved faster). The various estimates shown in Figure 1–1 suggests that the United States had about 1,200 robots in 1974 and perhaps 2,400 by 1977. By the end of 1981, it appears that there were at least 4,700 industrial robots in use in the United States, as compared to 14,200 in Japan, and a world total of almost 27,000 (Robot Institute of America, 1982).

Stories in the press also have reported that the numbers of orders for robots placed in the past year, in the United States, about equals the number of robots already in use. Industry and trade journals over the last year have consistently reported an "explosion" in orders for industrial robots, and a further indicator of suddenly increased interest is the appearance of major feature articles in several national news magazines about industrial robots, within the past year.[1] Bache, Inc., the investment firm, reports that U.S. sales of robots are projected to grow at a compound annual rate of 35 percent, from $100 million in 1980 to $500 million in 1985, to over $2 billion by 1990 (Conigliaro 1981). A more recent forecast reprinted in the RIA worldwide survey is even more optimistic. The survey reports that growth for total robot market in the United States is projected to reach $900 million by 1985 (15,000 units) and increase to $7 billion by 1990 (100,000 units). General Motors, with 300 robots installed as of January 1981 and approximately 1,600 installed as of mid-1982, has announced plans to increase that number to 14,000 by 1990 (General Motors 1982). In March 1982 GM announced a joint venture with Fijitsu Fanuc Ltd. of Japan to form a U.S.-based joint venture company to design, manufacture, and sell robotic systems. General Electric Corp. had 70 robots in its plants in 1979 and 200 by the end of 1981. Now General Electric has entered the

---

[1] For example, *Newsweek,* 30 June 1980; *Business Week,* 9 June 1980; *Time,* 8 December 1980; *Next,* May/June, 1980; *Discovery,* 1980; *GEO,* May 1981.

Figure 1–1.   Estimates of the U.S. Robot Population: 1970-81.

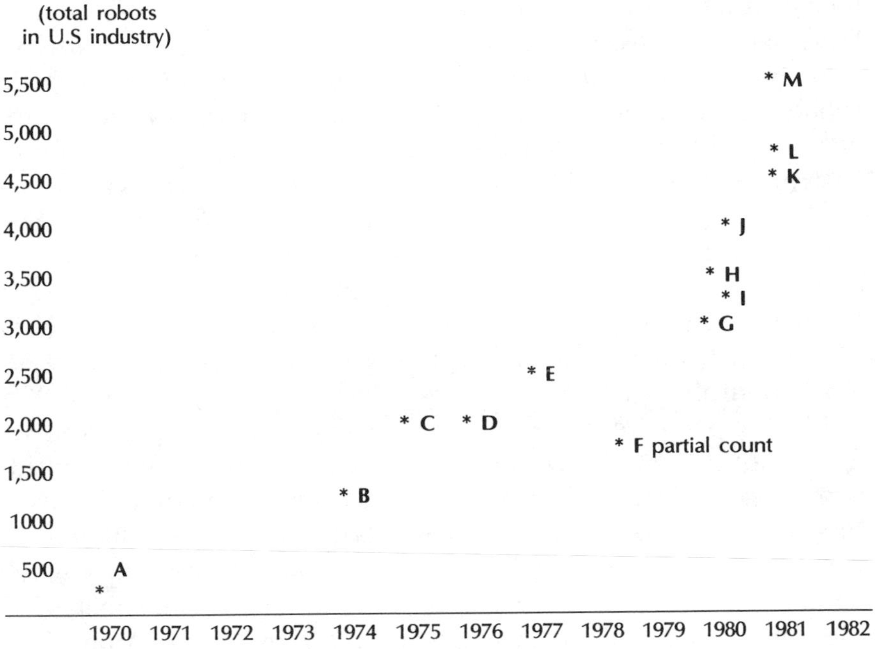

(total robots
in U.S industry)

5,500											* M	
5,000												
										* L	
4,500											* K	
4,000										* J		
3,500										* H		
									* I		
3,000										* G		
2,500								* E				
2,000						* C	* D					
								* F partial count			
1,500												
				* B							
1000												
500	A											
*											

1970 1971 1972 1973 1974 1975 1976 1977 1978 1979 1980 1981 1982

Point	Number of Robots	Date	Source
A	200	1970 (April)	Engelberger (1971).
B	1,200	1974 (Dec.)	Frost and Sullivan, US Industrial Robot Market, 1974.
C	2,000	1975 (Dec.)	Frost and Sullivan, The Industrial Robot Market in Europe, 1975.
D	2,000	1976 (Dec.)	Eikonix (1979).
E	2,400	1977 (Dec.)	Eikonix (1979).
F	1,600	1978 (Dec.)	American Machinist (1978).
G	3,000	1980 (Jan.)	Walt Weisel, Prab Conveyors.
H	3,500	1980 (June)	Business Week (1980), verified by Cincinnati Milacron.
I	3,200	1980 (Dec.)	General Motors Technical Staff (Bache, Shields estimate).
J	4,000	1980 (Dec.)	Walt Weisel, Prab Conveyors.
K	4,370	1981 (Dec.)	Aron (1981).
L	4,700	1981 (Dec.)	Robot Institute of America (1982).
M	5,500	1981 (Dec.)	Seiko Inc. Marketing Dept.

robotics field as a manufacturer (and its own largest customer) by licensing the rights to produce Italian and Japanese robots. Westinghouse Electric Corp. is also manufacturing its own robots and has also licensed rights to manufacture foreign-made robots. IBM has also recently started selling Japanese made robots and will soon market their own assembly robot. Both Texas Instruments and Digital Equipment Corp. are also widely regarded as potential new entries to the field (Business Week 1980). Bendix Corp., now the number two manufacturer of machine tools in the United States (following its acquisition of Warner and Swasey) has also begun manufacturing robots. New robotics firms or divisions of existing firms are being formed almost monthly.

Despite a ten-year head start over Japan, which produced its first robot in 1969 (based on a U.S. license), Japan is now considerably ahead of the United States in both production and use.[2] Based on the U.S. definition of a robot, Japanese production in 1980 was 3,200 units worth $180 million versus 1269 units ($100 million) in the United States. In fact, as of 31 December 1980 there were 11,256 installed operating units in Japan (Aron 1981) and over 14,200 by the end of 1981 (RIA, 1982). There are nearly 140 robot producers in Japan. In 1981, Paul Aron reported that the Japanese Industrial Robot Association (JIRA) forecasted output to increase from 3,200 units in 1980 to 32,000 in 1985 and 57,500 units in 1990 (based on U.S. definitions). Updated JIRA forecasts reprinted in the 1982 RIA worldwide survey project over 100,000 units in use by 1985 and over 325,000 in use by 1990! Growth after 1985 is expected to shift toward the more sophisticated "intelligent" models.

France has also made a national commitment to robotics. Starting in the early 1970s there are now twenty government-supported research centers in the field, and the Mitterand government has promised $2 billion to further robotics research. Twelve French firms are said to be manufacturing robots (including Renault, the auto manufacturer). The USSR did not produce its first robot until 1971-72. Yet, according to the Robot Institute of America survey, and to researchers who have visited Russia, there

---

[2] See Ayres, Miller, and Lynn (1981) and Lynn (1982) for a brief history of the introduction of robots into Japan.

are now 2,000-3,000 programmable units in place.[3] The Soviets expect a 50 percent per annum growth rate through 1985. Comparative robot population statistics are shown in Table 1–1.

Table 1–1. International Robot Population, February 1982.

Country	Type A	Type B	Type C	Type D	Total
Japan	—	6,899	—	7,347	14,246
United States	400	2,000	1,700	600	4,700
USSR	—	—	—	—	3,000
West Germany	290	830	200	100	1,420
Great Britain	356	223	54	80	713
Sweden	250	150	250	50	700
France	120	500			620
Italy	—	—	—	—	353[a]
Czechoslova-kia	150	50	100	30	330
Poland	60	115	15	50	240
Norway	20	50	120	20	210
Denmark	11	25	30	0	166
Finland	35	16	43	22	116
Australia			62		62
Netherlands	48	3	5	0	56
Switzerland	10	40	—	—	50
Belgium	22	20	0	0	42
Yugoslavia	2	3	5	0	10
Total	1,774	10,924	2,584	8,299	26,924

Type A: programmable, servocontrolled, continuous path; type B: programmable, servocontrolled, point-to-point; type C: programmable, nonservo robots for general-purpose use; type D: programmable, nonservo robots for diecasting and molding machines; type E: mechanical Transfer Devices (pick and place)—not shown.

[a] Population data for 1980. See Industrial Robot (1981).
Source: Robot Institute of America (1982).

[3] Professor Delbert Tesar, of the University of Florida, toured Russian scientific institutes in 1981. In his trip report, he cites researchers at the Leningrad Polytechnic Institute and at the Railway Institute (Leningrad) for these figures (Tesar 1981).

Forecasts made a few years ago that seemed excessively optimistic at the time now appear conservative. In 1980, prior to the
recent announcements by GE, GM, and IBM, Bache, the investment firm, projected 23,000 robots in use by 1990, assuming the
present robot manufacturers continue to lead the field. However,
Bache suggested that, if the large computer firms mentioned
above enter the market, volume production, reduced prices, and
heavy marketing activity could lead to as many as 200,000 robots in service by 1990 (Business Week 1980). By now, this forecast seems relatively modest, if anything. The existing potential
market for robots in the United States alone has been estimated
at over 400,000 units. Given improved capabilities, and, of
course, lower costs, the potential industrial robot market in the
United States would be much larger still. By one estimate, if the
cost of a $50,000 (1981 prices) robot could be reduced to $10,000
by 1990, demand might rise perhaps as high as 200,000 units *per
year* (Froehlich 1981). Thus, it seems that market penetration to
date has been both low and slow. The reason, very briefly, is
that industrial robots of the current vintage are not yet as cost-
effective as humans in most jobs. Even where robots are clearly
superior, as in spot-welding, the margin of superiority is still
fairly modest.

THE CONCEPT OF TECHNOLOGY ASSESSMENT

When a technology is just emerging into commercial application,
an attempt to anticipate the potential impacts on society is always fraught with difficulties. The potential effects are highly
sensitive to the rate and timing of adoption, as well as the level of
utilization. This is especially true for a decentralized technology
such as robotics and computer-aided manufacturing (CAM).
Long-range economic, technical, and social trends, as well as
chance events, will perturb the analysis both by affecting the rate
of diffusion of the technology and by reinforcing, or counteract-
ing, those events which have been forecast.

Until very recently there has been little public awareness of the
developing technology of industrial robots, little or no government involvement with the course of its development (in the

United States), and little thought as to what levers the government may have over development. Demand for policy intervention can be expected if it appears that the technology may impose significant social costs or if it appears to promise significant social benefits that will not be realized through the normal working of the marketplace. Advance notice of, and public discussion of, such possible impacts may promote a reasoned discussion that will avoid an emotional crisis response if adverse impacts do become obvious.

Policy analysis that attempts to anticipate and evaluate the possible future impacts of an emerging technology, in order to inform public policy formulation, is known as technology assessment. Technology assessors use a variety of analytical techniques, including trend extrapolation, historical analogy, delphi exercises with participation by experts and by potentially affected parties, and other aides to combining analytical reasoning, informed judgment, and creative imagination. A conceptual model of the process of technological development and diffusion is a useful tool in guiding the search for and the evaluation of potential impacts.

In the case of industrial robots, a useful conceptual model is one which focuses on the "user institutions," in this case the manufacturing segment of the economy, and in particular capital goods manufacturers, who will adopt and utilize the technology. The schematic model shown in Figure 1-2 suggests several steps in the process by which a host institution, or user, adopts and utilizes the technology, with resulting consequences for the user, other institutions or organizations with which the user interacts, the physical and social environment, and society at large. This model appears useful in tracing the consequences of a technology that is introduced incrementally, a few units at a time, in a number of like institutions, and operates at first by the side of the technology for which it will substitute.

The model pictures a process in which the user institution first adopts the new technology in order to perform some activity or function more effectively or more efficiently. The new technology is not an exact replacement for the older technology (in this case, human operatives) and does not have the same characteristics. In order to use the new technology most efficiently, that is, to cap-

Figure 1–2.   Impact of Technology Innovation on User Innovations.

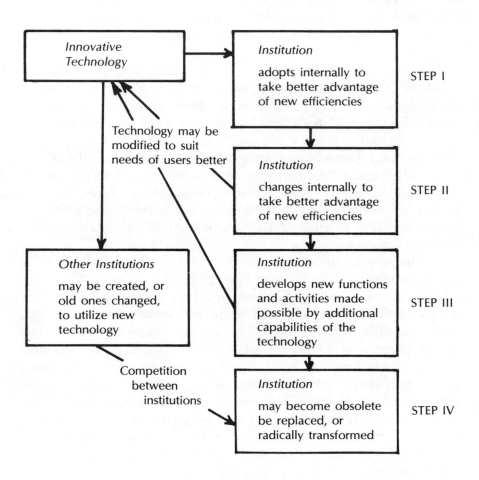

Source: Coates (1977).

ture all of its benefits, the institution will make internal changes or accommodations, such as modification of the management structure, redesign of the workplace, or changes in the product itself.

Usually, however, the new technology can also do things of which the older technology was not capable. For example, early office computers could "keep books" and handle payrolls more effectively than humans with adding machines and ledgers, but they also made possible more complex operations, such as on-line inventory control. Robots may be capable of working in hostile environments which humans cannot tolerate. The user tends to develop new activities, functions, or objectives that will make use of more of the capabilities of the technology. Each of these waves of change and adjustment sets up pulses of secondary and indirect changes, or impacts, in the institutions' environments and in the surrounding social structure.

In terms of this conceptual model, applied to industrial robots, we suggest that stage one is the use of robots as a direct replacement for human workers in existing jobs, in the existing workplace. Stage two would occur as the manufacturer begins to redesign the production process to better match robot capabilities with tasks. Stage three would occur when manufacturers, having available a more flexible production process, begin to put more emphasis on "quick response" to changing market requirements—thus shortening the product life cycle and accelerating the pace of technological change. Robots are now evolving very rapidly, and it is not clear what capabilities they may have (inherently or by later design) that are different in kind from human capabilities and that manufacturers may eventually seek to utilize. Already, however, robots have been utilized in the space program, so a possibility that comes immediately to mind is the use of robots to manufacture unique products in the gravity-free environment of space. Similarly, underwater robots may be the key to exploitation of vast resources in and under the ocean.

It is not within the scope of this book to attempt a systematic technological forecast either of technological advances in robotics or of the future rate and pattern of diffusion. Rather, we will take some estimates from the current literature and ask "what if?"—

that is, what is likely to occur if these estimates are correct. Our approach here is broad-brush, with the emphasis on suggesting reasonable possibilities and raising questions that should be addressed in a systematic inquiry.[4]

REFERENCES

American Machinist. The 12th American Machinist Inventory of Metalworking Equipment, 1976–78. *American Machinist* 122(12):133–148, December 1978.

Aron, Paul. *Robots Revisited: One Year Later.* Technical Report 25, Daiwa Securities of America, Inc., 28 July 1981.

Ayres, Robert U., Leonard Lynn, and Steven M. Miller. Technology transfer in Robotics: U.S./Japan. In *Proceedings of the First U.S.-Japan Technological Exchange Symposium.* Japan-America Society, 1302 18th Street, N.W., Washington, D.C. 20036, 1981.

Robots Join the Work Force. *Business Week, 9 June 1980.*

Capek, Karel. *RUR-Rossem's Universal Robots.* Oxford University Press, London, 1973.

Carnegie-Mellon University. *The Impacts of Robotics on the Workforce and Workplace.* Department of Engineering and Public Policy, Carnegie-Mellon University, Pittsburgh, Pa., 1981. A student project cosponsored by the Department of Engineering and Public Policy, the School of Urban and Public Affairs, and the College of Humanities and Social Sciences.

---

[4] Two technology assessments, both considered exploratory and conducted before recent developments in robotics, are available. The first, "A Preliminary Technology Assessment of the Introduction of Industrial Robots in the Motor Vehicle Industry" was prepared by Edward S. Murphy (1976), a graduate student in the Science, Technology, and Public Policy Program at the George Washington University, under the direction of Vary T. and Joseph F. Coates in 1976. The second, "Technology Assessment:The Impact of Robots," was carried out by Eikonix Corporation (1979) under a grant from the National Science Foundation. Several preliminary assessments have been recently published. In June 1981 a one-semester student project jointly sponsored by the Department of Engineering and Public Policy, the Graduate School of Urban and Public Affairs, and the Department of Social Science at Carnegie-Mellon University (1981) resulted in a report entitled "The Impacts of Robotics on the Workforce and Workplace." It considered the issue of labor displacement in detail and also discussed human resource programs, labor policies, and industry experience with introducing robots. In July 1981 the Office of Technology Assessment (1981) held a "Robotics Workshop" where several commissioned papers were presented to members of congressional staffs. The topics were applications of robots, growth of the robot industry, and potential social impact issues. OTA is carrying out a study of the impacts of robotics during 1982-83.

Coates, Joseph. *Telecommunications Policy* 1:196–206, 1977.

Conigliaro, Laura. *Robotics Presentation Institutional Investor's Conference: 28 May 1981.* Technical Report, Bache Halsey Stuart Shields, Inc., 19 June 1981.

Eikonix Corporation. *Technology Assessment: The Impact of Robots.* Technical Report, Eikonix Corp., September 1979. Sponsored by NSF Grant ERS-76-00637.

Joseph F. Engelberger. Economics and Sociological Impact of Industrial Robots. In *First National Symposium of Industrial Robots.* Illinois Institute of Technology Research, April, 1971.

Froelich, L. Robots to the Rescue? *Datamation*:81–96, January 1981.

1982 General Motors Public Interest Report. Technical Report, General Motors, April 1982.

Hogan, James P. *The Genesis Machine.* Ballantine Books, New York, 1979.

The State of Robotics in Italy at the Start of the 80's. *Industrial Robot* 8(3):176–177, September 1981.

Lynn, Leonard. *Two Challenges from Japanese Robotics.* Technical Report, Department of Social Science. Carnegie-Mellon University, June 1982.

Malone, R. *The Robot Book.* Jove Publications, New York, 1978.

Murphy, Edward S. A Preliminary Technology Assessment of the Introduction of Robots in the Motor Vehicle Industry. Master's thesis, George Washington University, Science Policy Program, 1976.

Office of Technology Assessment. *Exploratory Workshop on the Social Impact of Robotics: Summary and Issues.* Technical Report, Office of Technology Assessment, March 1982. Backround papers presented at an OTA workshop held in July 1981.

Reichardt, Jasia. *Robots: Fact, Faction and Prediction.* Penguin Books, 1978.

Robot Institute of America. *Robot Institute of America Worldwide Robotics Survey and Directory.* Society of Manufacturing Engineers, Dearborn, MI. 1982.

Tesar, Delbert. *Trip Report: Visits to Major Research Centers in Robotics in Europe and Asia.* Technical Report, Center for Intelligent Machines and Robotics, Department of Mechanical Engineering, University of Florida, 24 July 1981.

2 ROBOT/CAM TECHNOLOGY

ROBOTICS AND THE INDUSTRIAL REVOLUTION

For purposes of the technological discussion to follow some general remarks may be helpful in putting the present situation into historical perspective. The development of technological civilization may not be, in itself, evidence of "progress" from any moral or ethical point of view. But cumulative, technological change is a fact, as reflected in man's growing ability to affect and shape the natural environment and to extend it.

This manipulative capability of humans is attributable to a combination of a large forebrain and a versatile hand, with opposable thumb, freed for numerous activities by painfully achieved upright posture. The human forebrain seems to provide at least two capabilities that are rare or nonexistent in other animals. First, the big visual cortex—10 percent of the brain mass—permits the use of stereoscopic vision (rather than smell) as the primary sense, by creating an internal three-dimensional image (in color) of the external world and of the body in relation to that external world. This, in turn, creates the possibility of very complex learned physical (motor) skills.[1] Second, the forebrain contains "signal processing" capabilities of a high order—an obvious prerequisite for the development of language and communication. Language, in turn, permits articulation, accumulation, and transfer of information from individual to individual. It is also the start of symbolic manipulation and eventually mathematics and science.

The most significant technological innovations since prehistoric times have mostly been amplifiers or extensions of either the hand, the foot, or the eye-brain combination.[2] Until quite recently, in historical terms, these extensions were essentially mechanical—simple tools, vehicles (wheeled or floating) and weapons. When human muscles were inadequate, larger stronger animals were tamed and exploited. Later, wind and falling water provided some supplementary energy available to do work in modest amounts. Wood, leather, clay, bronze, or hand-smelted iron sufficed for the construction of almost all the artifacts associated with the preindustrial phases of technological development. Hotter fires and advances in iron-working technology made the so-called industrial revolution (1760-1830) possible.

The great "growth industry" of the industrial revolution was cotton textiles. It was rapidly rising demand for cheap cloth that created a demand for mechanization of spinning and weaving. Mechanical looms, in turn, required "prime movers" to provide power for them. This need encouraged James Watt and Oliver Boulton to develop the primitive steam engine of Savery and Newcomen into a practical source of stationary power for factories as the nineteenth century began. To manufacture machines, of course, required metals and metalworking tools, notably the lathe and milling machine. The development of the primary metals (no-

---

[1] It is noteworthy that virtually all animals are cerebrally "preprogrammed" as regards physical capabilities. A kitten is as good at playing with a ball or a mouse after a few weeks as a full grown cat. Not so with humans.

[2] Indeed, one of the direct precursors of modern robots were the articulated artificial hands made for amputees since the German knight, Goetz von Berlichinger, received the first iron hand in 1509 (Reichardt 1978). Before the end of the sixteenth century the military surgeon Pare was providing metal arms with flexing elbows and a hand that could openand close. Not much more could be achieved in this field until the advent of plastics and electronics made possible modern prostheses such as the Collins hand, 1961, the Ernst hand, 1961, the Roehampton arms, the Edinburgh arm, 1963-76, the Boston arm, 1969, and the Stanford arm, 1970.

tably, steel) and metalworking industries were thus essential concomitants of industrialization. The metal lathe has had virtually its modern form since about 1800 when Henry Maudslay created it complete with that crucial device, the V-slide, for advancing a tool toward the work with ease and accuracy. The milling machine was invented by Eli Whitney (of cotton gin fame) in 1818 as part of his project to manufacture rifles for the U.S. Army. It was tools such as these that provided the wherewithal for mechanization. However, it was not until after 1870 that large-scale production of machine tools made them widely available. One event that helped make mass production of machines possible was the development of the "automat" by Christopher Spencer. The automat was a programmable lathe used for producing screws, nuts, gears, and other stock, items the availability of which provided a tremendous boost to the usefulness and economic stature of the machine tool industry. The automat was "programmed" by fitting strip cams to the perimeter of a revolving drum. These guides would operate levers in sequence, moving tools to appropriate positions for machining the desired part. Cams were the dominant method of control for machine tools until the 1950s.

The automat was not the first programmable machine. As early as the 1720s Bouchon and Falcon in Lyons had designed looms for weaving patterns in silk which were partially controlled by punched cards or paper tape. These designs were improved successively by Jacques de Vaucanson in the 1770s and finally by Joseph Jacquard. In 1805 Jacquard introduced a practical automatic loom, 10,000 copies of which were in use in France alone within seven years. The method of "programming" that was so successful for the automatic loom was almost immediately adopted by Charles Babbage, the English mathematician, for his mechanical calculating machines, the "difference engine," 1823 and the unfinished "analytical engine," 1833-71.

While Babbage did not succeed in developing a viable mechanical calculating machine, his work was influential, and he was, in some sense, the "inventor" of the modern computer. In fact, the punched card retained its key role as a data input and storage device until very recently. Babbage tried to build a computer without either the necessary precision components or a sufficient demand for high speed mechanically assisted computation. These things came together only in the 1940s.

The punched card was the ancestor of digital data storage, but most machines (until very recently) have utilized analog controls because they lacked the capacity to rapidly process large volumes of digital data. Punched cards occupy the same place in the history of digital controls as cams have in the history of analog controls.

Numerical control became practical because of the development of the high-speed computer.[3] In 1934 Konrad Zuse, a German construction engineer, began work on program-controlled computing machines. He completed a fully operational, general purpose computer in 1941. All of his work was destroyed during World War II. Howard Aiken built an electromechanical machine, the automatic sequence-controlled calculator at Harvard (for IBM) during 1940-42. The ENIAC, the first truly electronic machine, was built at the University of Pennsylvania during the period 1943-48 by J. Presper Eckert and J.W. Mauchly, under contract to the U.S. Army. The EDSAC at Cambridge University was the first computer with a stored program (1949). Meanwhile, Eckert and Mauchly founded a company to exploit the new technology, and accepted a contract to build Univac 1 for the U.S. Census Bureau in time for the 1950 census. The fledgling firm delivered the product 18 months late (in 1951) and ran into severe financial problems, but still managed to be first in the market. IBM unveiled its first commercial machine, the 701, in 1952.

The most significant invention in the history of computers was the transistor, developed by John Bardeen, Walter Brattain, and William Shockley in 1948 at Bell Telephone Laboratories. Commercial production began almost immediately and led to the rise of a number of important new firms, notably Texas Instruments, Inc. William Shockley himself moved to Stanford University and created Shockley Transistor, Inc. This firm was not immediately a great financial success—eventually it was absorbed by Fairchild Camera and Instruments—but Fairchild, in turn, spawned half a dozen other firms (sometimes known as "Fairchildren") which themselves split and resulted in the remarkable phenomenon of integrated circuits and "Silicon Valley." The greatest success story is that of Intel Corp., formed by Gordon Moore and Robert Noyce about 1961. Intel was the developer of the microprocessor

---

[3] See Ralston, 1976 for a brief history of digital computers.

("computer on a chip") which is revolutionizing the control of machines of all kinds—including robots.

Numerical control (NC) technology involved special purpose electronic controllers that could be "programmed" with a sequence of operations specified in terms of a "workpiece" coordinate system and translated (via software) into a total coordinate system. The tool itself was controlled by means of "servo" mechanisms. Interfaces between NC tools and outside sources of workpiece specifications became more sophisticated over time. At first, in the era of direct numerical control, the specifications had to be programmed off-line in "machine language" on tape by a specially trained programmer. Each machine had its own dialect. The machine was controlled on-line by the tape reader. Later more general-purpose languages were developed, and in the mid-1960s Cincinnati Milacron Corporation introduced a system for controlling a number of machines directly from a central computer through the satellite controllers on each machine. With the CNC, or computer numerical control, each tool is controlled by a (digital) minicomputer, which may or may not be tied to a central computer. The hardware is general purpose, but each machine tool has its own unique software.

Servo-control technology is based on the notion of continuously comparing a stream of data about the "real" state of an object with its desired state. The comparison and correction constitutes a "feedback loop." The concept goes back at least as far as James Watt's steam engine speed governor, but it has become truly practical in recent decades as electronic means of sensing and transmitting information evolved. Modern servo-control technology was initially developed for artillery fire-control purposes in World War II and for missile-control purposes immediately thereafter. The center of this military work was the MIT Instrumentation Laboratory—now called the Charles Stark Draper Laboratory after its founder. Some of the key mathematical theory came from work by Claude Shannon and Norbert Weiner, both at MIT. Weiner popularized the term "cybernetic," which embraces all kinds of feedback-control processes, including those applicable to humans as well as robots.

The first practical general-purpose apparatus to record signals magnetically and play them back to control the motions of a mechanical device seems to have been invented by George Devol in

1946 (U.S. patent 2,590,091, issued in 1952). This patent was first licensed by Remington Rand Corporation for use in connection with the Univac computer but was found to be too slow for the purpose, and rights were later returned to Devol. Later, Remington-Rand did develop and market the UMAC general-purpose process controller in 1964. UMAC was not itself a great success, but general-purpose controllers later became widespread in industry.

We now come to robots.[4] A working robot is basically a programmable multipurpose manipulator. It requires the confluence of sophisticated mechanical technology (the arm, wrist, end-effector, and actuators) and sophisticated control technologies. As will be emphasized later, it was only when microprocessors were integrated into the control systems that industrial robots became applicable to a wide range of industrial tasks.

One of the earliest industrial manipulators was a motorized rotary crane with a gripper for removing ingots from a furnace, developed in 1892 by Babbitt (U.S. patent 484,870). An extension of Babbitt's device, and one of the direct precursors of the modern robot, was the "teleoperator" or remote-controlled arm developed in the 1940s and 1950s to handle dangerous radioactive material. Key developments have been attributed to Goertz (U.S. patent 2,695,715; 1954) and Bergsland (U.S. patent 2,861,701; 1958).

The Atomic Energy Commission (AEC) contracted with Hughes Aircraft Company to build a sophisticated two-armed mobile remote-control manipulator or "teleoperator" to carry out experiments with radioactive materials in an AEC facility at Albuquerque. The "mobot," built in 1960, had vision (television), hearing (microphones at the wrists), and "soft" hands, but it was not designed to do the same task over and over from memory. Another line of precursors is traceable to the prosthetic arms mentioned earlier.

A more direct line of ancestry to present-day applications can be established to a series of paint-spraying machines developed in the late 1930s, for example, Pollard (U.S. patent 2,286,571; 1942) and Roselund (U.S. patent 2,344,108; 1944). These devices could be programmed by leading a jointed arm through a desired path. This created a magnetic signal which was recorded and could be

---

[4] See Ayres, Miller, and Lynn (1981) and Ayres and Miller (1982) for chronologies of the development of robot technology.

"played back" to guide the arm. Roselund's machine recorded the signal on a grooved disk similar to a phonograph record.

The first commercial robot was introduced by the Planet Corporation in 1959. It was (and is) a "pick-and-place" device programmed and controlled by limit switches and cams. A competing unit was introduced by American Machine and Foundry (AMF) and acquired by Prab Conveyor Company. The earlier Versatran robots were non-servo-controlled point-to-point devices operating on independent axes. Such robots are still quite widely used for simple repetitive motions. Meanwhile George Devol's control patents were bought as a package by Consolidated Diesel Engine Company (CONDEC), which set up a subsidiary, Unimation Inc., to develop the new technology.[5] Unimates, which appeared on the market by 1960, were first to utilize servo systems, and to provide a general path-control option. In effect, Unimation introduced numerical control (NC) as applied to robots. Unimation recently held about 40 percent of the U.S. market.

Early NC robots had computer-like functions, such as memory, but were made up of electronic logic components "hardwired" to perform a specific set of tasks. Electronic controls were used to essentially duplicate the functions of other "hard automated" control functions. Robots actually controlled by digital computers were developed in the early 1970s. The first mini computer controlled robot was the T3, commercialized in 1974 by Cincinnati Milicron Corporation, the leading machine tool manufacturer in the United States. Microprocessor controlled robots followed several years later. The computer-controlled or "softwired" robot is far more powerful than a machine with specialized electronic logic circuits. It can work in several coordinate systems, be programmed "off-line," interface with sensors, and so on. Computer-controlled robots are now becoming very specialized peripheral features of general-purpose computers. Yet as of January 1980 there were only around 3,500 robots installed in the United States out of an existing potential market that has been estimated at over 100,000 units.

Although the third industrial revolution—based on electronics and computers—is still mostly in the future, there are growing in-

---

[5] The name Unimation is derived from the expression "universal automation," as used by Devol in a 1954 patent application to describe his first programmable transfer device.

dications that it has finally begun in earnest.[6] While computers are not yet able to operate many kinds of machines, and the superiority of computer control is still limited to relatively special cases, the recent trends are all in this direction. Computer hardware costs (especially of microprocessors) are dropping rapidly, general purpose computer software is improving, and much of the necessary infrastructure—programmers and engineers familiar with computer concepts—is in place. By the same token the real cost of semiskilled human labor (to industry) is rising sharply, and its quality is probably diminishing. Moreover, indirect costs of human labor, in terms of associated health, safety and environmental requirements are increasing rapidly. This combination of circumstances has created strong incentives for industry to begin to replace semiskilled human machine operators of various kinds by programmable devices on a larger and larger scale. The increasing use of robots is part of this overall trend.

TECHNICAL LIMITS OF ROBOTS

How far will the current "robot boom" go? At the present time, it is clear that numerical controls and programmable robots are established with human workers only in a fairly small number of simple, repetitive jobs that can be done by a single arm with two fingers. Now that the capabilities of industrial robots are becoming better known, it is evident that the mainstream of manufacturing jobs is just beginning to become subject to programmable automation. What the commercial robot of 1980 still lacks (at least, in adequate degree) is a high-quality feedback-control system: an ability to sense its own position and motion in relation to the working environment and to utilize this information to modi-

---

[6] For our purposes, we label the first industrial revolution as the period between 1760 and 1830 when a number of major inventions were applied to iron-making, to steam power, and to the mechanization of spinning and weaving, especially of cotton. We label the second industrial revolution as the period between 1870 and 1910, during which the Bessemer steel production process, electrification, the telephone, the compact internal combustion engine, and mass production methods were developed. We choose 1970 as the beginning of the third industrial revolution since it marks the beginning of mass production of microprocessors. See Ayres (1982) for more extensive discussion of the implications of the third industrial revolution.

fy its actions in real time—as humans do—so as to compensate for variations and irregularities in that environment. It also lacks the ability to coordinate two hands with many fingers. On reflection, it is evident that both sensory and interpretive (or cognitive) skills are required to do these things.

A major problem to overcome before robots can achieve their full potential is the lack of artificial sensory systems with a resolution comparable to the human eye-brain combination. Artificial senses may, to be sure, utilize parts of the electromagnetic spectrum or the audio spectrum (e.g., SONAR) not directly accessible to humans. But it is not enough for a machine to be able to send and receive signals. The received information must then be interpreted. In effect, the automation must be able to construct a consistent internal "picture" of itself as it moves in, and modifies, its environment. Thus, perhaps the weakest link in robot technology is the primitive state of interpretive or cognitive abilities. It is therefore not surprising that robot research today is focusing in the area of sensory systems and artificial intelligence. The problem may merely be one of inadequate computer power. A black and white television camera constructs a picture from 512×512 "pixels," a stream of 1×10^7 data bits per second. The vertebrate retina, by contrast, has a million individual light sensors and "preprocessing" to determine edges, and curvature and motion are done by 20 million neurons. However, the full-color three-dimensional image produced in the human visual cortex requires a thousand times more capacity than the retina alone and about 10^5 times more capacity than current commercial vision systems (Moravec 1978). If computer processing power continues to become cheaper at the historical rate—a factor of 10 every five to seven years—robot vision should eventually be competitive with human eyesight, but that time is still several decades away. The following description from the abstract in Moravec's (1981) Ph.D. thesis illustrates the present state of the art of "autonomous" visually guided robot rovers:

> The Stanford AI Lab cart is a card-table sized mobile robot controlled remotely through a radio link, and equipped with a TV camera and transmitter. A computer has been programmed to drive the cart through cluttered indoor and outdoor spaces, gaining its knowledge from images broadcast by the onboard TV system.

The cart uses several kinds of stereo to locate objects around in 3D and to deduce its own motion. It plans an obstacle avoiding path to a desired destination on the basis of a model built with this information. The plan changes as the cart perceives new obstacles on its journey.

The system is reliable for short runs, but slow. The cart moves one meter every ten to fifteen minutes, in lurches. After rolling a meter it stops, takes some pictures, and thinks about them for a long time. Then it plans a new path, executes a little of it, and pauses again.

The program has successfully driven the cart through several 20 meter indoor courses(each taking about five hours) complex enough to necessitate three or four avoiding swerves. A less successful outdoor run, in which the cart skirted two obstacles, but collided with a third, was also done.

Harsh lighting (very bright surfaces next to very dark shadows) giving poor pictures and movement of shadows during the cart's creeping progress were major reasons for the poorer outdoor performance.

Even long-range potential in this area may be quite limited. There is no doubt that where humans excel in relation to mechanisms (including computers) is in the rapid coordination of physical actions in response to a quickly changing set of perceptions about the environment. To put the long-range potential of robots in perspective, we have compiled a set of lists of increasingly difficult tasks (Table 2–1). Those on the third list may be possible for robots, but probably not within two or three decades to come. Those in the fourth list may require many decades.

The differences between jobs a robot can easily do and jobs a robot may never be able to do (column 4) are not a matter of precision or complexity. A job that is very hard for current robots is one that (1) cannot be preprogrammed, (2) involves decisions that must be continuously modified in real time, as circumstances (or information regarding them) change—includig two-handed coordination—and such that (3) the decision criteria inherently cannot be reduced to a set of mathematical equations. The last point is one that deserves particular attention because nowadays many people who are not professionally involved with computers tend to assume that virtually any problem that is quantifiable can be reduced to equations and "solved.' This is absolutely not so. A computer model can solve the equations of motion for a set of

Table 2–1   Tasks that Robots Can Do and May Be Able To Do.

Things Present (or Past) Robots Can Do	Things Next Generation Robots Will Be Able To Do	Things a Very Sophisticated Future Robot May Be Able To Do	Things No Robot Will Ever Be Able To Do (Probably)
Play the piano	Vacuum a rug (avoiding obstructions)	Set a table	Cut a diamond
Load/unload CNC machine tools	Load/Unload a glass blowing or cutting machine	Clear a table	Polish an opal
Load/Unload die casting machines, hammer forging machines, molding machines, etc.	Assemble large and/or complex parts, TV's refrigerators, air conditioners, microwave ovens, toasters, automobiles	Juggle balls	Peel a grape
Spray paint on an assembly line	Operate woodworking machines	Load a dishwasher	Repair a broken chair or dish
Cut cloth with a laser	Walk on two legs	Unload a dishwasher	Darn a hole in a sock/sweater
Make molds	Sheer sheep	Weld a cracked casting/forging	Play tennis or pingpong at championship level
Deburr sand castings	Wash windows	Make a bed	Catch a football or a frisbee at championship level
Manipulate tools such as welding guns, drills, etc.	Scrape barnacles from a ship's hull	Locate and repair leaks inside a tank or pipe	Pole vault
Assemble simple mechanical and electrical parts: small electric motors, pumps, tranformers, radios, tape recorders	Sandblast a wall	Pick a Lock	Dance a ballet
		Knit a sweater	Ride a bicycle in traffic[a]
		Make needlepoint design	Drive a car in traffic[a]
		Make lace	Tree surgery
		Grease a continuous mining machine or similar piece of equipment	Rapair a damaged picture
		Tune up a car	Assemble the skeleton of a dinosaur[b]
		Make a forging die from metal powder	Cut hair stylishly
		Load, operate and unload a sewing machine	Apply makeup artistically
		Lay bricks in a straight line	Set a multiple fracture
		Change a tire	Remove an appendix
		Operate a tractor, plow or harvester over a flat field	Play the violin[c]
		Pump gasoline	Carve wood or marble
		Repair a simple puncture	Build a stone wall
		Pick fruit	Paint a picture with a brush
		Do somersaults	Sandblast a cathedral
		Walk a tightrope	Make/repair leaded glass windows
		Dance in a chorus line	Deliver a baby
		Cook hamburgers in a fast-food restaurant	Cut and trim meat
			Kiss sensuously

[a] Assuming the other vehicles are not robot controlled.
[b] Admittedly a computer could provide very valuable assistance.
[c] But it could "synthesize" violin music.

planets or a ballistic missile in a gravitational field or even, conceivably, a mechanical juggler. But the game of tennis is too complex for a formal mathematical model, and will be for the foreseeable future.

A tennis player, to function effectively, must "see" the position, velocity, and acceleration of his own and his opponent's torso, legs, arms, and hands (with fifty or more degrees of freedom) in relation to the court and the air. He must also see, hear and respond to the trajectory and spin of the ball, the way it comes off the racket, and how it bounces. After integrating this voluminous data, he must decide on an appropriate action at each moment; that is, where and how to hit the ball and where to go himself in order to be ready for the next shot.

The kinds of information required in tennis might seem comparable (allowing some poetic license) with the data needed to fly a jet aircraft or a rocket shuttle. These are tasks humans are quite good at, yet robots can also do them. In fact, the space shuttle will, in normal circumstances, be controlled by computers rather than human pilots during reentry and landing. The reasons illustrate one of the potential strengths of robots in certain situations. The human pilot has only one pair of eyes and one pair of hands. Consequently, he can receive only primary sensory data from the shuttle's sensors in serial fashion, on one channel at a time (disregarding auditory and tactile inputs), and he can only manually issue new commands to the vehicle at a correspondingly limited rate.

A robot, by contrast, can be provided with as many simultaneous data channels for receiving and sending data as necessary, plus a parallel processing capability. In the case of a space shuttle during reentry, it is virtually essential that the pilot should possess multichannel monitoring and parallel processing capability. The physical attributes of the vehicle—being an artifact—can, of course, be "built in" to the decisionmaking model used by the (robot) pilot.

In the case of tennis, on the other hand, a single channel of visual data input is essentially sufficient (in addition to tactile and proprioceptive inputs needed for body control). The human brain is superior to the computer-controlled robot—at least for the present and the foreseeable future—in sensory and cognitive ability. The tennis player constructs an internal mental "model" of the situation at each moment. The model, of course, incorporates the physical attributes of his body, his opponent, the racket, the ball, and the court. From this model, he projects (i.e., simulates) an expected situation a few seconds into the future, decides on an optimum response, and issues appropriate instructions to the muscles. He continually modifies these expectations as the situation changes, of course, and previous instructions are correspondingly reinforced or altered.

Why couldn't a robot be designed to perform equally well? At present, as we noted above, the best optical scanners or "image processors" (e.g., video cameras) are far inferior in quality to the human eye coupled with the visual cortex. Improvements will occur in time, but only by lavish use of data processing to extract significance from the "raw" sensory data—much as the human brain does with the signals it receives through the eyes and ears. Thus, the vision module for a robot tennis player would require computer power far greater than the largest computer in existence (the Cray), or even contemplated. Another more fundamental reason is that the simulation and decisionmaking "model" in the human tennis player's brain is the result of a very long and tedious learning process involving tens of thousands of hours of practice with actual legs, arms, and hands, as well as hundreds of hours of practice with racket and ball (not to mention millions of years of evolution). There is no set of mathematical equations to describe the kinematics of the human body—it is simply too complex. How, then, would the tennis-playing robot be pre-programmed? We return to this question in a moment.

Of course the motions of a robot—like a rocket—may be much simpler (more rigid, fewer degrees of freedom) than a human body, and consequently, its motions may be more easily described in an explicit mathematical form. But even if we postulate a mathematically describable system, the associated optimal control problems—that is, the problem of making an optimal choice for each of many independent variables, subject to variable constraints based on information about the state of the environment varying in real time—is far beyond the present—or foreseeable—capabilities of mathematics (even with unlimited computational capability). In short, the formal mathematics of tennis are vastly more complex than the mathematics of a space shuttle.

Lacking computable mathematical solutions of the equations of motion, the only other possible approach to designing a robot tennis player is to design a robot brain capable of self-programming (that is, learning through practice as a human does). In fact, this is true of all tasks listed in column 4 of Table 2–1. This statement does not imply that robots will never be capable of learning. Nor does it imply that each robot will have to learn for itself, as humans do. The real advantage of robots, in the long run, lies in the fact that once the knowledge is acquired—however tediously—it can be transmitted to the next robot very easily indeed. Each skill need only to be learned the hard way once. Knowledge transfer (between robots) thereafter becomes almost trivial. The importance of designing robots capable of learning by experience is, therefore, immense. In principle, one may easily suppose that a computer could be taught to learn from experience, that is, to modify its own programs. But, as a practical matter, we do not know how to program a computer to solve very general classes of problems (including the problem of self-programming). The development of artificial intelligence (of necessity) proceeds from the particular to the general. Thus, computers can be taught to solve relatively narrowly defined "types" of problems. Clues are derived from a variety of sources, including cognitive psychology, behavioral psychology, neurophysiology, and even anthropology. But progress in this field consists in continuously generalizing the solvable problem types. What we do not really know, at present, is how far we have come and how far we have yet to go before computers can be designed to learn by themselves. Indeed, we don't even know for certain whether it is possible to program a computer (of familiar logical structure) to learn from experience in the way humans do. Clearly, it would be foolhardy to suggest that this problem cannot eventually be solved. (Obviously, biological organisms can learn from experience. Why not computers?) But it would be equally premature to suggest that a robot in the year 2000 (or 2050) will have a true learning capability, comparable in any way to that of humans. By the same token, once the nature of the human learning process is sufficiently well understood, it may be possible to vastly speed up the human learning process, also. We expect computers to transfer their accumulated knowledge by reproducing the contents of their memory cells, in another computer. Why should not some analogous procedure be possible for humans? If the brain is essentially a mechanism, vastly accelerated learning techniques should indeed be possible—just as the slow natural process of building up antibodies to disease can be accelerated by inoculation with artificially attenuated disease organisms.

ROBOT TECHNOLOGY: A REVIEW

We have already defined industrial robots, loosely, as "manipulators." More precisely, industrial robots are essentially programmable multijointed arms (with grippers or tool holders at the end) capable of moving a tool or workpiece to a prespecified sequence of points, or along a specified path within the arm's reach and transmitting precisely defined forces or torques to those points. As such, it is more appropriate to think of a robot as a mechanical 'operative," rather than as a machine tool of the familiar kind (Paul and Nof 1979).

The following review of robot technology is intentionally brief, since more detailed discussions of the technical characteristics of current robots are readily available, especially in Engelberger (1980), Toepperwein and Blackman (1980), Japan Industrial Robot Association (1982), Gevarter (1982), Jablonowski (1982), and Fisher and Nof (1982). In the next few pages we describe the major categories of robots under several headings, namely, architecture, actuation and performance, controls and programmability, and sensors.

Architecture

Regarding kinematic and structural design, there are four general architectural types distinguishable in terms of coordinate systems applicable to the arm(s):

- Cartesian (rectilinear) (x, y, z); examples: Advanced Robotics Cyro 750, Seiko 400, Thermwood Cartesian 5, Westinghouse Series 7000
- Cylindrical (r, z, ω); examples: Copperweld CR-10, General Numeric GN1, Prab FA
- Spherical (r, ψ, ω); examples: Bendix Robotics AA160, General Electric MH33, Unimate 2000

- Revolute (polar articulated, or "jointed") (w, ψ, ω); examples: ASEA IRb-60, Cincinnatti Milacron T3, DeVilbiss TR-3500, Unimation Puma 550

The four types are schematically illustrated in Figure 2–1. Each of these has 3 degrees of freedom, sufficient for the arm to reach any point within a volume of space defined by the maximum extension of the arm. Work volumes have very different shapes, as shown.

Of these types, the anthropomorphic revolute or "polar articulated" architecture, requiring only cylindrical (rotoidal) couplings, offers comparatively large working volume at low cost and good ability to avoid obstacles along the positioning path. The chief disadvantage of polar articulated architecture has been that servo controls for continuous path operation are more sophisticated than some other architectures. However, recent advances in computer processing power (and cost) have effectively eliminated this drawback. For this reason, Cartesian (rectilinear) and cylindrical architectures are likely to assume reduced importance in the future except where exceptional positional accuracy is needed. As 3 degrees of freedom are required to reach any point within the working volume, 3 additional degrees of freedom are required to deliver the tool or workpiece in any arbitrary orientation. This may not be necessary in some cases—for example, if the workpiece is itself cylindrical or spherical—but most robots have a "wrist" articulated more or less as illustrated in Figure 2–2.

Actuation and Performance

Among the key performance characteristics (other than control and programmability) are the following:

- Payload
- Reach
- Speed
- Accuracy and repeatability

Figure 2–1 Architecture and Work Volume: Cartesian, Cylindrical and Spherical.

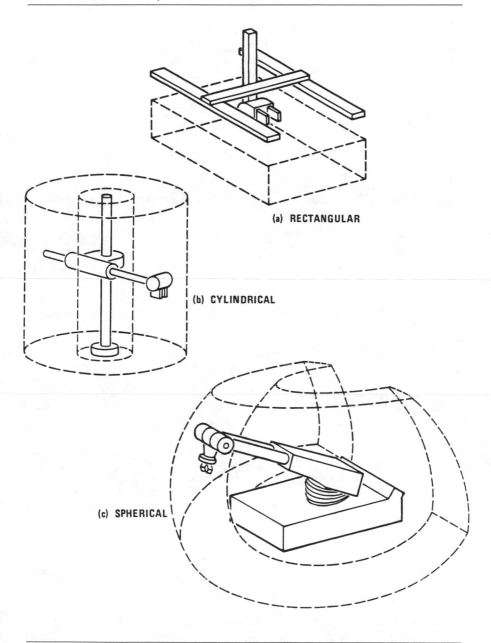

(a) RECTANGULAR

(b) CYLINDRICAL

(c) SPHERICAL

Source: Toepperwein and Blackman (1980).

Figure 2–2 Architecture and Work Volume: Revolute.

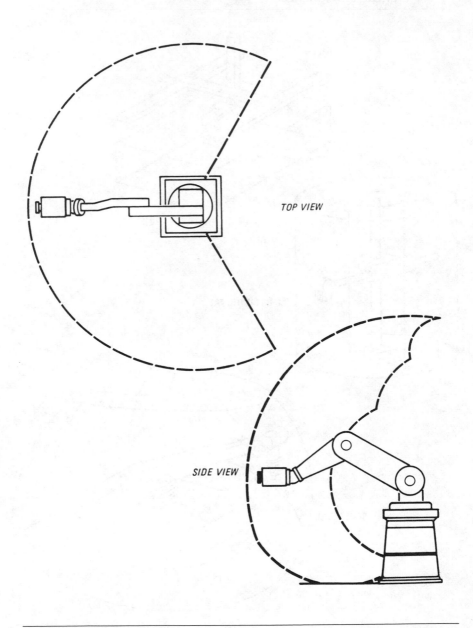

Source: Toepperwein and Blackman (1980).

Not surprisingly, there are important performance tradeoffs among these. That is, one can increase one only at the expense of others. Hence, a different combination will be best for a given purpose, and no single robot design can be optimal for all purposes.

Specifically, large payloads are incompatible with long reach (due to mechanical disadvantages), high speed (because of inertial problems), and high accuracy and repeatability (for both reasons). Long reach is similarly incompatible with high speed. Indeed, for most geometries, the maximum payload varies between different parts of the working volume (inversely with reach).

Three kinds of actuators are generally available: pneumatic (compressed air driven), hydraulic, and electric. In the first two cases power is transmitted to the joint as pressure. This pressure is converted to torque and kinetic energy by a simple motor (actually a pump acting in reverse). Pneumatic actuators are the simpler and cheaper of the two but cannot handle as heavy loads as high speeds as the hydraulic systems. Pneumatic drives are generally used for small, inexpensive point-to-point "pick-and-place" robots, such as the Seiko.

Conventional electric robots are complicated by the use of mechanical transmissions such as gear trains, screws, belts, or chains. This adds weight and cost. The main advantage is thought to be superior accuracy and eased controllability. However, recent developments in digital encoding permit servo-controlled systems for hydraulic units to achieve the accuracy and repeatability associated with conventional electric drive, but with greater load-carrying capacity (Toepperwein and Blackman 1980). On the other hand, a new type of direct drive robot developed at Carnegie-Mellon University uses lightweight electric motors (based on the use of high-strength rare earth permanent magnets) embodied in the joints themselves. This innovation permits drastically improved performance, especially in terms of arm speed and accuracy (Asada and Kanade 1981).

So far the word "accuracy" has been used loosely. However, it is important to distinguish three slightly different concepts operationally in relation to an arbitrary never-before approached "target" position:

- Spatial resolution is an irreducible uncertainty of final tool position arising from the inability of the robot to detect or to control physical motions smaller than a certain limit. Operationally, the resolution is defined by a circle around the target position, within which the tool tip will be found when the target is manipulated (by an operator via the control system) to approach the target as closely as possible.

- Repeatability is a measure of the uncertainty in final position when the initial position actually achieved is recorded and the arm is moved away and then commanded to return to the recorded position. Here the robot is operating in "'tape recorder mode." Repeated attempts will define a circle somewhat larger than that due to resolution alone, due to errors and uncertainties in the recording and playback systems. There will be long run deterioration due to joint wear, oil leakage, and so on.

- Accuracy is a composite measure of uncertainty that applies in the case where the target is defined in an independent external coordinate system and the computer-controller must calculate all the robot joint-positions in terms of the robot's internal coordinate system. Such calculations are always inaccurate to some degree. This situation arises whenever the "training" situation is simplified in some respect from the "operational" situation. For instance, the robot may be trained to work on a stationary workpiece, but the operational case may involve workpieces moving along a transfer line.

All three uncertainty concepts—spatial resolution, repeatability, and accuracy—can be defined in terms of a linear distance, such as the radius of a circle. The relationship between the three is such that each circle is contained within the next, such that the radius of resolution < radius of repeatability < radius of accuracy. It must be emphasized that, strictly speaking, the third measure does not apply to a robot that is not computer controlled.

Controls

The simplest (nonrobotic) machines execute one instruction at a time, usually at the command of an "on-off" switch. The operator decides when the switch should be turned on, observes the progress of the work, and—based on a continuing flow of information processed through his eyes and interpreted by his brain— decides when to turn the switch off. The above procedure is characteristic of mechanization. It is exemplified by the operation of a vacuum cleaner, a power lawn mower, a lathe, or a power drill. The situation is so familiar that most people do not realize its inherent complexity.

To make the jump from mechanization to automation, a new feature must be added. It is the idea of preprogramming a sequence of distinct operations without intermediate intervention by the human operator. (This concept was first implemented in practice in the textile industry by the Jacquard loom in 1801.) It took a long time to apply the concept to a sequence of motions along (or around) several axes independently, with changes of instruction (e.g., start, accelerate, decelerate, stop) triggered by the arrival of the moving part at a prespecified location on each axis. The earliest "pick-and-place" robots included both of these features but with instructions conveyed by means of cams and limit switches.

The next step in sophistication is coordinated motion along or around several axes simultaneously, and the introduction of continuous feedback (i.e., servo) control. Servo control makes it unnecessary for the operator to decide in advance exactly where (on each axis of motion) there should be a change of instruction. Instead, the machine monitors its own position and velocity continuously with an "ideal" trajectory that is stored in the machine's memory. The ideal trajectory can be "taught" to the robot in several different ways, which we discuss shortly. The robot has a control algorithm that tries to minimize the positional and velocity differences, either as the machine approaches its end-point target (as in the case of point-to-point robots) or along the whole path.

The former is often sufficient for operations such as machine loading/unloading or spot welding, whereas the latter is required for spray painting or arc-welding. There are intermediate cases where only the end points need be specified precisely but the intermediate path should be somewhat constrained, for example, to avoid an obstacle or to maximize effective speed and/or minimize power requirements. The last situation arises because when the robot arm is freely extended its inertia—here the load on joints and actuators—is much greater than when the arm is pulled in. (The analogous situation can be observed when a figure skater goes into a spin; he begins rotating slowly with arms and perhaps a leg extended, and speeds up by drawing the arms as close as possible to the body. The same phenomenon can be seen when a diver goes into a "tuck" position to speed up the rate of spin.) This can be done roughly, without specifying the entire path in detail, by specifying one or two intermediate or "via" points, as illustrated in Figure 2–3. There are three common methods of teaching or programming a robot at present: manual, lead-through, and walk-through. These terms are perhaps not self-explanatory. In a manual system, the controls themselves are mechanical, as in most machine tools. Controls are set "off-line" by presetting cams on a drum, connecting hydraulic lines, or setting limit switches on each axis of motion. This type of programming is typical of a variable sequence pick-and-place manipulator, such as the Autoplace, or Seiko units.

Where more complex motions are required, electronic memories and servo controls are also indicated. These are similar to numerical controls (NC) for machine tools. To create a set of instructions for such a robot, one can hold a small control unit and push buttons which direct the arm along the desired path, while recording a number of intermediate points. This procedure is "lead-through." Or one can physically move the arm over the desired path—known as "walk-through," again recording a number of intermediate points. The electronic controls may be special-purpose (analog) or general-purpose (digital) hardware. In the latter case, specialization to a particular task is accomplished by means of programmable software. The robot is controlled by a minicomputer.

The next step in sophistication, currently under development, would be for the instructions governing path motion to be generated off-line from a higher level in the factory control hierarchy. Such instructions could eventually be based on a computer simulation of the robot-machine tool system, as in CAD. A hypothetical future "intelligent" robot may conceivably be given a capability to develop its own operational strategy—or sequence of actions—by trial-and-error testing outcomes in terms of generalized, functional, or goal-oriented instructions. Such a robot would remain largely independent of higher level controllers. However, as mentioned earlier, a true learning capability analogous to the way humans learn by experience seems fairly remote at present.[7]

---

[7] Research in the field of artificial intelligence is actively pursuing this goal (Fox 1981a; Carbonell, Mickalski, and Mitchell, 1982). See Chapter 8 for further discussion of machine learning.

Figure 2–3   Use of Via Points to Speed Up Manipulator Motions.

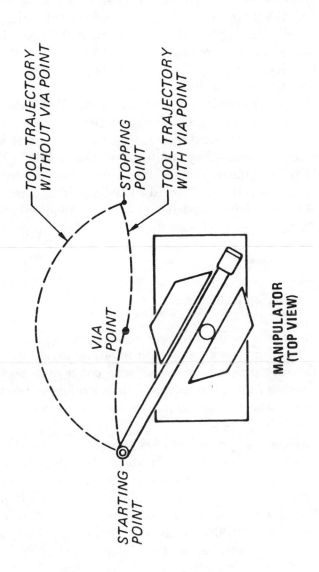

Source: Toepperwein and Blackman (1980).

The most fundamental limitation of present-day robot capabilities is the need for pre-specification of the task in complete detail. Human workers can often be given vague orders, and (based on a lifetime's worth of experience) stillknow just what must be done. Given the instruction, "Put the nut on the bolt," any normal child could accomplish the task without further ado. In contrast, today's robot could not do this task with so few instructions. Every aspect of the operation, including how to hold the bolt and the nut and how to engage them, has to be specified in detail. Robot programming languages are being developed to simplify the task of transmitting instructions to the manipulator. The languages are classified in terms of the amount of knowledge and reasoning power they require of the robot. "Explicitly-programmed" languages require the user to specify manipulator positions and trajectories. "World-modeling" languages in the future will use very simple instructions merely to specify what is to happen. Manipulator positions and trajectories are generated automatically.

Sensing and Learning

Most tasks in the real world cannot be prespecified to the required degree of precision, and require adjustments and modifications as the task proceeds. Such adjustment is only possible with help from sensory data inputs. Real time, on-line adjustments are needed for a variety of different reasons, including:

1. The workpieces are not always oriented in exactly the same way as they enter the work station or they have variable shapes (e.g., animal carcasses or saw logs), or they are not perfectly rigid (e.g., pieces of rubber, leather or fabric).
2. The operation to be performed must be modified, depending on a variable attribute of the workpiece (e.g., the location of a burr or defect in metal or the conformation of grain patterns in wood).[8]

---

[8] The reasons robots are not suitable for tasks such as gem cutting and polishing is that the size and location of defects generally determine the strategy of cutting. Problem-solving or "intelligent" computer programs which could generate a plan for cutting the gem

3. The operation to be performed depends on unpredictable factors in the environment (e.g., steering around an obstruction on the factory floor or avoiding other workpieces, tools, etc.).
4. The operation to be performed requires coordination of several limbs and/or balancing forces in an unstable equilibrium (e.g., riding a bicycle, walking on stilts, juggling).

The first two items on the list encompass categories of problems that are common in industry, and require too much sensing, decisionmaking, and adaptation to be performed by the current generation of industrial robots. Picking standard parts from a bin is trivially easy for humans and exceedingly difficult for a robot.[9] The same is true for cutting logs or fitting pieces of cloth together. To perform such tasks at all, sensory feedback systems are essential. The robot must sense the appropriate attributes of the workpieces as the operation proceeds and make corrective maneuvers as needed. The robot must also recognize when the desired result has been achieved and end the operation. Alternatively, it must recognize when the workpiece has been damaged and should be removed from the line. These are major challenges to the present state of the art.

The third and fourth categories define problems that at present can only be solved in highly simplified environments. Solutions that can successfully cope with the complexities of the real world are beyond the likely capabilities of industrial robots for years, if not decades, to come. A robot may be taught to steer around an obstruction on a factory floor, but it is unlikely that a robot could soon be made to recognize and cope successfully with all the contingencies that might be faced by the driver of a car, for instance. Beyond this, the high degree of eye-brain-body coordination and heuristic decisionmaking required to ride a bicycle in traffic seem unlikely to be duplicated by any commercial robot for many decades. Humans, of course, do this sort of thing extremely well.[10]

---

would first have to be developed before the task could be efficiently performed with a robot.

[9] The first experimental system using a robot to pick randomly oriented cylinders out of a bin was demonstrated in 1981 (Kelly 1982).

[10] As as example of the state of the art in walking robots, Fredkin and associates at MIT have designed a robot with a single jointed leg that theoretically should be capable of doing somersaults! But this is not really indicative of a high level of physical skill. It

It is convenient to distinguish three categories of robot sensing, following Raibert (1981):

- Internal
- Remote (e.g., vision)
- Contact (e.g., force, pressure)

The first category applies to the internal monitoring of the robot. The second and third relate the robot to its working environment.[11]

Vision has received the most research attention to date of all the remote sensing techniques. Vision systems that distinguish silhouettes in a black and white image have been in use since 1973. These systems depend on lighting and on other techniques to produce a high contrast binary image (Agin 1980). They can be trained or programmed to recognize shapes. There are several commercially available "binary" vision systems used to inspect, count, locate, and orient parts, as well as to guide (servo) a manipulator to an object in real time. More advanced vision systems of this type that have the capability to distinguish many shades of gray are approaching commercialization. A more powerful remote sensing technique still under development is "structured light." In this approach, light is projected onto an object in a precise pattern. The range to the object is determined by triangulation, and the object's three-dimensional shape is inferred from the loci of intersection of the beam with the object. Tactile sensing means, in effect, detecting pressure changes, that is, by means of piezo electric cells arranged in a grid. A state-of-the-art sensor being developed at MIT has a grid of 16 × 16 pressure cells. Torque sensing involves the detection of rotational drag (or load) as when a generator is attached to a turning shaft. Other senses can obviously be important in special cases. Temperature sensing could be important to a robot designed to handle hot (or cold) parts, for instance, although the most likely form of temperature

---

happens that the equations of motion in this case can be solved exactly. Marc Raibert has built a one-legged "dynamically stable walking machine" that hops and balances in 2 degrees of freedom (Raibert 1981). Hirofumi Miura at the University of Tokyo has built a two-legged walking machine which is dynamically stable in 3 degrees of freedom.

[11] See Kinnucan (1981) for a discussion of current research in robotic sensing systems.

sensing at a distance would be an optical color sensor. Ferromagnetic or diamagnetic sensing could also be utilized by robots working with metal objects.

While the range of possible sensory inputs is quite large, the interpretation of the sensory signals by the robot's controlling "brain" remains a separate problem. In effect, the sensors provide a stream of input data in one language, but the information cannot be used for purposes of decisionmaking until it is internally translated and comprehended. This is the problem of cognition. Learning capability relates to the creation and modification of an instructional program, on-line, based on a goals statement and sensory input data. A software interface, not yet developed, will be needed to achieve effective sensory feedback and learning by experience. This area of fundamental research (artificial intelligence) lies squarely in the domain of computer science. As noted already, it is not at all clear how soon robots will acquire learning capabilities enabling them to self-program. It appears, however, that such capabilities would require five or six orders of magnitude more computer power than is now available.

APPLICATIONS IN STANDARD INDUSTRIAL TASKS

Tasks that robots are capable of doing may be classified conveniently as follows (de Gregorio 1980):

1. *Pure displacement*
   a. Load/unload machines
   b. Parts manipulation (e.g., packaging, sorting and conveying, orientation)
   c. Palletizing
2. *Displacement and processing*
   a. Spot welding
   b. Continuous welding
   c. Mechanical and electrical parts assembly
   d. Electronic parts assembly
   e. Spray painting
   f. Cabling
   g. Cutting

    h. Other processing operations with portable tools
3. *Displacement and inspection* (subdivision of load/unload category by types of machines or operations)
    a. Machine tool: deburring, drilling, grinding, routing
    b. Plastic materials forming and injection molding
    c. Metal die casting
    d. Hot forging and stamping
    e. Cold forging
    f. Cold sheet stamping
    g. Furnace load/unload
    h. Heat treatment
    i. Foundry: sand casting, investment casting

Robots have initially had the greatest success to date in welding applications, followed by machine tool loading/unloading, foundary work, and spray painting, as seen in Table 2–2. Even in these established applications many practical problems remain to be solved.

The introduction of robots into most hot-metal-forming activities came about initially because of safety or quality control issues. For example, the workshop conditions in the drop forging industry are recognized as some of the most arduous faced by human workers. Heavy preforms need to be retrieved from high-temperature furnaces, manipulated into the jaws of a heavy, fast-

Table 2–2   Distribution of U.S. Robot Population by Application.

Application	Estimated Percentage of United States Robot Population Within Application
Welding	34%
Machine loading/unloading	20%
Foundry work	19%
Other applications	13%
Painting and finishing	12%
Assembly	2%

Source: RIA (1982).

moving press, and perhaps rearranged in the press for subsequent forging strikes before removal and stacking. These tasks are inherently dangerous and tiring. It was in this area that teleoperators (remote-control manipulators) were used for many years before programmable robots of the current type took over. The teleoperator removed the man some distance from the forge interface, but at the cost of reduced flexibility and sensitivity of control. In fact, in order to fully exploit the industrial robot, it is advisable to make some further changes to the typical forging operation. As emphasized before, existing robots are insensate and cannot look onto a red-hot furnace and identify parts that are randomly distributed. A vision (or other sensory) system to solve this problem will have to be capable of operating reliably in an environment of high temperatures (up to 450°K), oil mist, and oil vapor (Schraft, Shults, and Nicolaisen 1980). The conservative approach is therefore to install an induction-heating furnace (rather than an air circulating furnace) where parts enter on a conveyor, are heated discretely, and are then available in some known orientation for a robot to retrieve them.

Similarly, some features of the forging dies themselves may need to be redesigned. It is quite common in forging to find that the component sticks to the upper or lower die after the strike has been made. A human operator can easily see where the part is at any given moment, retrieve it, and manipulate it for subsequent strikes. Without vision sensing, a robot cannot do this and even with modern vision capabilities, as seen in research laboratories, the speed of coordination between vision and gripper is relatively low. The best solution is for the forging dies themselves to be more carefully designed, so that the part remains in a known orientation after the first forging strike.

In addition, the workpiece changes its shape as the forging process proceeds. To cope with this problem, more flexible (or adjustable) grippers are needed for the robot in order that the various part shapes in the forging sequence may be manipulated and handled. Deburring of forged blanks, a tricky and dangerous operation, would seem to be an ideal application of robot. But burr edges are irregular, resulting in variable reaction forces, and the motions required to achieve a smooth surface by grinding are highly variable and complex. Optical and tactile feedback systems with adequate sensitivity for these tasks were not commercially available, at least until recently (Schraft, Shults, and Nicolaisen 1980).

All of the major industrial robot suppliers have implemented installations in forging, and it appears that, with some specified *ad hoc* engineering and careful design of furnaces and conveyors, both open-die and closed-die forging operations can be operated effectively by industrial robots. Of course, cost is the main factor in evaluating a new installation. To overcome the various technical limitations of robots, a new "forging cell" generally needs to be installed before existing robots can be exploited. For large forging industries, this would seem prohibitive, and it seems plausible that the human operator would be preferred for most applications, at least until more competent robots become available.

Even so, the older, experienced forging operators are reaching retirement age and it is difficult at the present time to attract younger men to such hazardous operations. Especially with new emphasis on "near net shape" forming, the forging industry is one that is expanding. It may well be that the lack of experienced manpower in this area will force the use of industrial robots in the forging industry.[12]

In spot welding, two surfaces are held together with electrodes, current flows through the adjacent materials, and the heat that is generated causes a molten pool to form, joining the two components. This process is very common in the auto industry. In arc welding, a flame and filler electrode follow the butt joint of two plates to be joined. As with spot welding, the joining takes place as these molten areas cool. Both kinds of welding operations require close control of the parameters. In the past, an experienced welder would choose the best angle of attack, the dwell time, or welding speed, and the supply voltages. In fact, the teach and repeat capability of modern industrial robots makes welding an early candidate for actual replacement of human workers.

The essential point is that the experience of the human welder is, in fact, stored in the weld sequence controller. In the case of spot welding, the weld sequence controller locates the spot, controls the pressure exerted by the C-clamp, turns the current on and off, and times these actions. In the case of arc welding, the

---

[12] This might be a driving motivation for the use of these technologies in a wide range of manufacturing applications. See Eaglen (1980) and Sutton (1980).

controller instructs the inert gas to be turned on, feeds the wire, turns on and off the power arm and controls all of the time sequences for these. The robot arm itself follows the continuous path that opens and closes, and carries the welding gun and the wire feed mechanism. The feed mechanism is connected to the weld controller, and the gas line for the gun also trails to the controller. In summary, if humans have been displaced from the welding operation, it is not so much industrial robots that have made this possible as the development of automatic welding controllers. In spot welding, at least, an insensate robot is sufficient to the purpose, and the current robot designs provide all the sophistication that is needed. For arc welding, however, visual feedback would be desirable to ensure that the robot arm does not stray from the desired path.

With regard to welding application—especially continuous seams—there are still many unsolved problems to automatic magazining and clamping of parts to be welded, and control over welding parameters (and quality), and precise positioning of seams. Existing applications are primarily spot welding in mass production lines, such as auto bodies. Other applications await improved sensory feedback, orienting, and magazining devices. Each application tends to be rather specialized, whence market penetration is still quite slow (Schraft, Shults, and Nicolaisen 1980).

Investment casting, based on the ancient "lost wax" process, has provided an interesting application for industrial robots. Briefly, the process consists of creating a hard wax model of the component to be made, coating this model with some temperature-resistant ceramic material, burning out the wax to leave a hollow core, and then using this hollow core as the casting mold for the object to be produced in metal. The process is typically used for gas turbine or rocket engine parts made of very expensive (nonmachinable) superalloys. The initial use of the wax creates a faithful replica, and it is interesting that this kind of casting technique spans the ages from Egyptian objets d'art to modern engineering components. In practice, the initial wax molds are made from master dies and then joined together in the form of a "tree." Essentially, this is a rack of wax components.

Traditionally, the human operator would pick up this tree and begin to swirl the molds around in a silica slurry to build up the ceramic coating. The slurry is a fine dispersion of refractory particles, and the refractory layer is gradually built up on the wax molds, with successive dipping, hardening, and drying processes. Gradually, a large shell mold is built up and then fired at temperatures above 1,000°C in order to provide strength. One key aspect is that the gradual building up of these shell molds means that the weight of the tree increases over time. Until industrial robots were introduced into this kind of operation, there was a lower limit as to how many molds could be placed on a tree. Thus, the use of robots with high weight-carrying capacity (up to 300 pounds) has increased the productivity of this process. In addition, the fact that human operators tend to vary their actions inadvertently with each dipping and swirling sequence has often meant that the molds vary in thickness from tree to tree and within individual trees. After melting out the wax and pouring in the metal, the final component's size can vary with shrinkage patterns in the mold. A uniform shell core is extremely desirable, inasmuch as a ruined casting is a total loss. As with welding operations, strong and highly repeatable, but insensate, robots are ideally suited for this kind of task.

Metal cutting machine tools can be loaded and unloaded by hand, by robots, or by integrated devices fed by automatic transfer lines, as in an automobile engine plant. The role of robots here will be limited to cases where automatic transfer lines are inappropriate (e.g., because a variety of different parts must be machined), but batch sizes are large enough to justify numerically controlled machine tools, fed by a robot. Moreover, because robots cannot (yet) handle nonoriented parts, the optimum present application is one where the robot unloads one machine and transfers the part to another machine. The operational linkage between robots and other machines is discussed in the next section.

It was pointed out earlier in this chapter that as late as 1977, 98 percent of the machine tools in the United States were still *not* numerically controlled (American Machinist 1978). Some are automatic, but most are not primary candidates for robot loading and unloading. On the other hand, the number of NC machine tools in use doubled from 1972 to 1977 and has certainly continued to increase rapidly. By 1982, the total could approach 100,000. Many of these NC machines could in principle be served by robots.

A vital task that is not presently an important application of robots, but may become so in the future, is parts assembly. With minor exceptions, assembly-line jobs cannot be efficiently accomplished by present-day robots for several reasons, including inability to recognize and pick up a desired part from a mixed collection, and lack of a sufficiently flexible multipurpose gripper. These limitations can be removed, to some extent, in newly designed plants where, for instance, all parts are palletized or otherwise preoriented as they are produced or enter from outside, and handled automatically thereafter.[13] The other, and more general, approach to the problem is to develop robots with sensory (e.g., vision) capabilities. The need for interface software between sensors and robot controls has already been noted. A high-level computer language is also badly needed to reduce what would otherwise be an extremely time consuming (and expensive) setup procedure for each assembly to tractable levels. An additional technical difficulty is the need in many assembly operations for two or more arms to work together in coordinated fashion. This can only be accomplished to a very limited extent with existing robots, as previously noted.

The applicability of robots to parts assembly may be quite limited for some time to come, however, even assuming technical advances in robot capabilities. Paul and Nof (1979) have compared, in detail, a simple water-pump assembly task involving positioning the pump top and a gasket over a base and inserting two screws, as performed by a robot and a human. A number of interesting points emerge from the differences in physical capabilities. The human has two hands with flexible fingers, tactile sense, and eyes. The robot has only one arm, simple gripper, limited compliance, and limited sensory ability. Nonetheless, the robot can do some things better than the human, including more precise positioning and orientation of the gasket and pump top, which negate the human sensory advantage in this case. The human is much better than the robot at selecting, picking up screws out of a bin, and orienting them for insertion—almost entirely by "feel." The robot must have a screw

---

[13] Westinghouse Electric Company has been working under NSF sponsorship to establish the feasibility of an automated batch assembly line for small electric motors. The plant would be capable of producing 450 different models, at a rate of 1 million units per year. Robots would be completely integrated into the proposed facility, with a labor saving of two-thirds, compared to a conventional plant (Abraham et al. 1977a Cowart et al. 1981).

feeder provided, with controlled orientation, and a positioning guide. But the robot is considerably more efficient at the actual screw insertion for two reasons: it has greater torque capacity than the human wrist (the human needs a torque wrench), and it has 360-degree rotation. The robot also "knows," from torque sensing, when to stop turning the screw driver.

It is clear from this example that a given assembly job would not, in general, be carried out by the same sequence of "work elements" by humans as compared to robots. Certainly the worker would carry out the task in a way that takes advantage of his natural human abilities. Since a robot has different abilities, it can be expected that the optimum means of accomplishing a task by a robot would involve major alterations in the choice and sequence of work elements. In short, each task must be restructured to exploit the capabilities of the robot. It can be determined only by extended experimentation (and perhaps computer simulation) whether any given task is more "efficient" for a robot or a human.

INTEGRATION OF ROBOTS INTO CAD/CAM SYSTEMS IN METALWORKING

The application of industrial robots in activities relating to "machining cells" is receiving considerable attention. The basic sequences performed within a "cell" are outlined in Figure 2–4. Raw materials flow through the factory, through preparatory functions (such as cleaning), through the "cell core" where they are machined, and on to supplementary functions (such as grinding and inspection). Planning functions such as design, tool engineering, programming, and scheduling "control" the flow of the products through the cell. The traditional method of metal part design has been to create a series of drawings on paper, possibly to construct a prototype, and then to make the production decisions that lead to the manufacture of a specific batch of components on a series of machine tools. The activities of the designer and those of the production planners have generally been disparate, and the manufacturing sequence has not, in the past, been "integrated." The recent developments in CAD/CAM systems and software are rapidly bridging this gap (Taramin 1980). Systems are now commercially available that integrate all manufacturing stages between designing parts at a video screen to generating the instructions to operate the machine tools (Machover and Blough 1980). Initially, a designer works at a computerized graphics terminal that displays a high-precision three-dimensional represen-

Figure 2–4   Product Flow in a Manufacturing Cell.

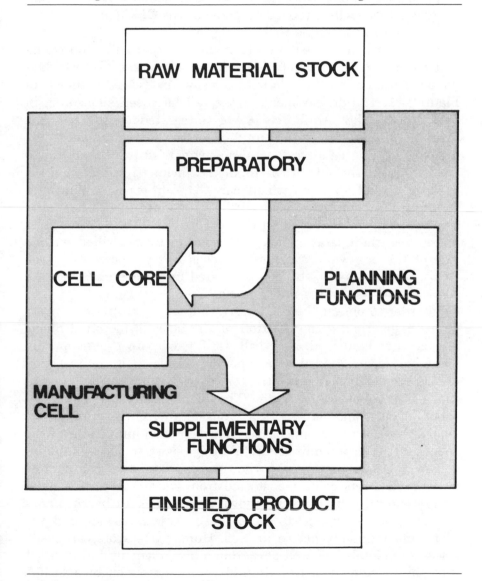

Source: Bjorke (1979).

tation of a component. The designer interactively modifies the component features and dimensions while viewing it from various perspectives. Following the design stage, decisions are made about the tooling to be used, and then the cutting paths can be superimposed on the design on the screen in order to verify the manufacturing sequences. Once this has been achieved, the system generates the machine tool commands for the CNC (computer numerical control) machine tool.

The recently installed robotic manufacturing cell for drilling and routing airplane components installed by the General Dynamics Corporation, Fort Worth Division, in Texas, is shown in Figure 2–5. The development of the cell has been sponsored, in part, by the U.S. Air Force ICAM (integrated computer aided manufacturing) program. Figure 2–5 shows the drilling and routing work on a large aircraft panel being carried out by an industrial robot. At the side of the robot (a Cincinnati-Milacron T^3) is a range of end effectors which carry the different drilling and routing tools. The panel being drilled is shown on the left-hand side of the figure, mounted on the rotable box. In the sequence of operations, the operator mounts the panel to be drilled on one side of the box while a second panel is being processed by the robot. Between parts, the box is rotated into place and then the robot begins its drilling and routing.

In order to obtain the necessary accuracy, a template that accurately determines the position of the holes is mounted above the panel to be drilled. The drill itself, carried in the gripper on the robot, is surrounded by a tapered bushing. It is the insertion of this tapered bushing into the template that improves the accuracy of the robot driller/routes. The installation of the industrial robot, jigging, and fixtures, was considerably cheaper to install than a five-axis machining center. The technical limitations of the robot system have required careful design work to successfully integrate the manipulator with the other parts of the manufacturing system. Parts being transferred from station to station must be presented to the robot in known orientation. In batch manufacturing operations, parts are typically transferred around the plant and randomly placed in bins. Human operators can easily select a randomly oriented part from a bin, orient it, and insert it into the machine tool (for example, into the chuck of a lathe). However, at the present time, basic design limitations of industri-

Figure 2–5   Part Drilled and Routed by Robot.

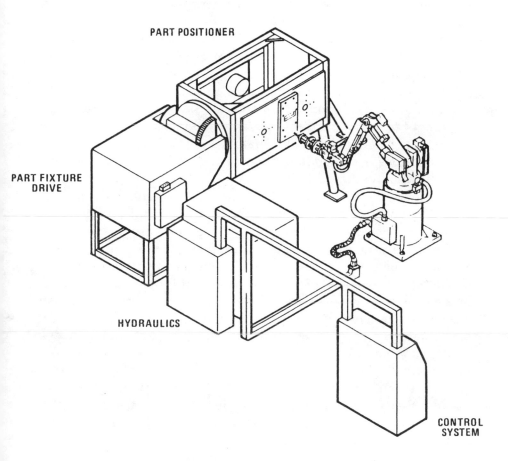

PART POSITIONER

PART FIXTURE
DRIVE

HYDRAULICS

CONTROL
SYSTEM

Source: Golden et al. (1980).

al robots require that parts be carefully stacked in known orientations and presented to the robot in such a way that the need for positive identification and reorientation is minimized.

In the next few years, we can expect to see industrial robots being installed in many medium-batch-size manufacturing plants, servicing two or three computer numerically controlled (CNC) machines. There will be a strong emphasis on the use of inexpensive microprocessors to coordinate the various pieces of hardware in such a cell. Stand-alone robots are crucial to the success of such systems, to load/unload the cutting or forming machines and to move parts from one machine to another (Figure 2–6). In addition to manipulation within the cell, there is a potential need for robots to carry out preprocessing functions, such as cutting raw bar stock and palletizing. There is also a need for supplementary functions, such as deburring, heat treating, surface plating, and (eventually) assembly.

In order to carry out a "closed loop machining operation" where the robot may also replace the routine metrology (measurement and inspection) operations in a manufacturing cell, tactile feedback is essential. While some dimensional measurement checks can be made on the machine tool itself with sensors placed in the tool changer, there will still be a need for measurement off-line. Such measurements are normally done at present by human inspectors. In moving toward fully automated cells, the robot will participate in this task via exact placement of parts in measuring stations. Technical developments that enable robots to be more versatile will clearly lead to more widespread installation in the manufacturing industry. For example, the development of a "universal gripper," or the ability to identify and pick up a part placed randomly on a moving conveyor or in a bin, is an important area of current research.

The next step in systems integration is for several machining cells to be linked together in a flexible computerized manufacturing system (FCMS), which further extends the the number of manufacturing steps that can be automated (Table 2–3). Workpieces with a given range of variability can be automatically subjected to differing production processes as necessary, by means of very sophisticated control and transport systems. The development of mini and microcomputers for control has made FCMS practical. Robots interact with numerically controlled machine

Figure 2–6   Technological "Core" of a Robot-Integrated Manufacturing Cell.

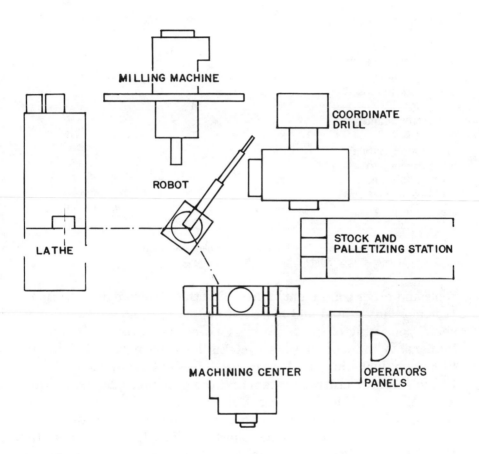

Source: Holmes (1979).

Table 2–3   Comparison of Manual Manufacturing Steps Elimination by Various Degrees of Computer Control.

| Step | Production methods | | | |
	Conventional	Stand alone NC	Machining center	FCMS
1. Move workpiece to machine	M	M	M	C
2. Load and affix workpiece on machine	M	M	M	C
3. Select and insert tool	M	M	C	C
4. Establish and set speeds	M	C	C	C
5. Control cutting	M	C	C	C
6. Sequence tools and motions	M	M	C	C
7. Unload part from machine	M	M	M	C

M = manual operation; C = computer-controller operation.
Source: General Accounting Office (1976: 38).

tools and other equipment, controlling the sequence of operations. It is more integrated and more automated than a traditional "job shop" consisting of machines in isolation, operated by individual humans. A number of such systems have been reported in the literature, including the Kearney & Trecker Corporation's (1978) Flexible Manufacturing System and the Cincinnati Milacron Variable Mission Manufacturing Centers installed in the United States, the East German "Prisma 2" System, the system at the Messerschmitt-Bolkow-Bolhm plant in West Germany, and the Fanuc and the Yamazaki systems in Japan. According to one recent estimate, Japan has installed nearly 30 Flexible Manufacturing Systems, while the United States, western Europe, and eastern Europe, including the Soviet Union, have installed around a dozen systems each. (Lerner 1981; Yoshikawa, Rothmill, and Hatuany 1981).

A relatively simple example is shown in Figure 2–7 representing a flexible three-cell system for making planetary pinion gears

Figure 2-7.   Flexible Computerized Manufacturing System.

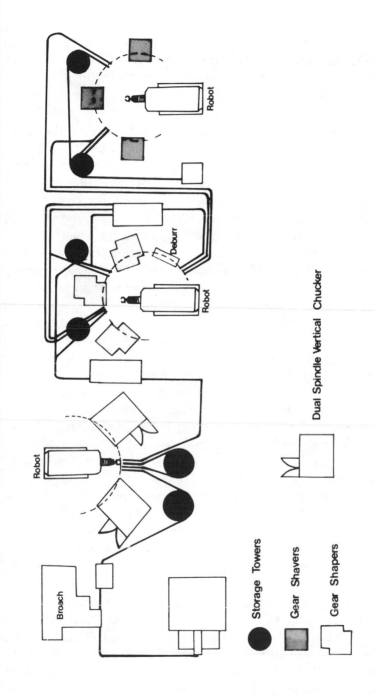

Source: Adapted from Jablonowski (1980).

at the Massey-Ferguson transmission and axle plant in Detroit. The manufacturer had originally planned for hard automation but found the use of robots (instead of a customized transfer line) to be both less expensive and quicker to install. This is a significant indicator of coming changes in manufacturing strategy.

The discussion so far has focused on applications of robots in metal cutting and forming operations in batch manufacturing, where programmability is an important attribute of the machine tools themselves and the interlinking assembly functions. It is interesting to note that metal cutting (i.e., machining) has become the dominant approach to metal shaping in most computer aided manufacturing (CAM) operations. Since machining involves significant raw-material wastage by chip formation, it is appropriate to analyze other "near-net-shape" metal forming methods, such as casting, forging, and powder forming, in terms of programmability. Unfortunately, at present, none of these rival forming technologies presently offer the same flexibility as does cutting, where the same set of tools can be programmed to produce a large number of components within a generalized family. Casting involves the manufacture of molds, while cold forging and powder metallurgy, for example, require precision dies—typically machined. This tends to make such processes uneconomic except for relatively large batch sizes.

Nevertheless, there is now an emerging interest in making alternative "near-net-shape" forming processes more programmable, because of their conservative use of materials and energy. Apart from this, there is a realization that many aspects of hot and cold forging die design are craft skills that preferably need to be quantified in some way as a prerequisite to embodiment in programmed instructions. Aspects of this kind of work are underway in the U.S. Air Force's ICAM program and its Process Modelling component. In these programs, both sheet metal forming and hot forging are being analyzed. In this research, information on material properties in the form of stress-strain curves, die-work friction effects, and elastic die deflections under load is being gathered and stored in a computerized data base, along with mathematical simulation models, for the benefit of the die designer working at an interactive computer terminal. While at present such systems are in their infancy and do not lead to the optimum die design, at least the trial-and-error aspect of die manufacture can be reduced in cost. Moreover, the interactive system is designed to "learn from its mistakes" in the sense that new data is fed into the data base.

An important recent technical development in die making is the use of electrodischarge machining (EDM) to shape the forming die. In the process, the electrode is graphite in the male shape corresponding to the desired female die. The key element of programmability can thus be introduced during the machining of this graphite electrode (usually by a ball-ended milling cutter on a four- or five-axis CNC machine). This facilitates a direct link between the designer and die manufacture, which in recent installations has dramatically reduced die costs.[14] This trend clearly suggests a prospect of using forgings for smaller batch sizes than heretofore.

The relevance of this to a discussion of robots is twofold. First, a flexible, programmable mold-making or die-making machine—utilizing feedback controls and learning capability—qualifies as a robot, according to most definitions, though it would not physically resemble the devices we have considered heretofore (Wright and Holtzer 1981). Second, there are applications for conventional pick-and-place or manipulative robots within a powder metallurgy cell (see Figure 2–8), as in a machining cell. For example, insertion of linear membrances, powder insertion, and compaction unloading and inspection are all tasks suitable for existing robots. It is worthy of note that the development of improved mold or die-making machines will tend to favor greater use of near-net-shape castings or forgings, as opposed to machining, in the future. Any change such as this will reduce the number of process steps in metalworking and, therefore, the cost of batch produced metal products—notably machine tools and capital goods themselves.

APPLICATIONS IN OTHER MANUFACTURING SECTORS

Robots are potentially applicable in manufacturing sectors other than metal processing, though a variety of special problems arise.

---

[14] Under the U.S. Air Force ICAM program, work is being done on sheet metal forming. In the work so far, simple strips, typically 12×1 inch, have been bent in a forming die consisting of spaced plungers. However, in this work, individual plunger positions are controlled by a computer interface prior to locking.

Figure 2–8   Schematic View of Computer-Robot Integrated Manufacturing Cell for Powder Metallurgy.

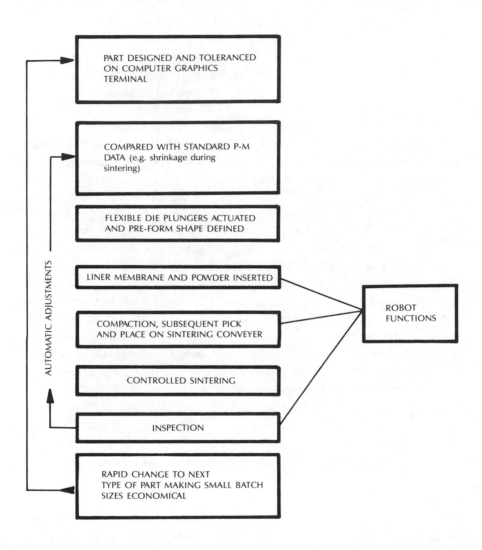

Source: Wright and Holzer (1981).

The following conclusions are based on a study by Schraft, Shults, and Nicolaisen (1980):

- *Leather processing.* Applications would probably be restricted to machine loading. Grippers would have to be modified.
- *Shoe manufacturing.* Possibly applicable to the loading and unloading of plastic molding machines.
- *Rubber processing.* Pattern recognition capability is a prerequisite for handling materials in the plastic stage prior to vulcanization. The most promising candidate for future automation is tire manufacture, but retrofitting existing work sites appears unfeasible.
- *Asbestos processing.* Automation of material handling operations, especially grinding, appears very appropriate for robots with current technology.
- *Plastics processing.* Operation of feeding calendering machines and vacuum forming machines is already highly automated. Robots can be used widely for loading and unloading injection molding machines, except where fiberglass reinforcing is used. In the latter case, temperature and other sensors would be required because of cycle time constraints. Application to pressing is being held back primarily by lack of automated presses. Spraying polyester resin coatings is difficult because the spray guns tend to clog and must be constantly rinsed. Foam spraying is also tricky because the foam will harden inside the gun if any interruption occurs. Special protocols and equipment will probably be needed to use robots in these cases, but it is only a matter of time before this application is developed.
- *Food processing.* Probably applicable primarily to materials handling, packaging, and inspection operations not specific to the product.
- *Clay and cement products manufacturing.* Not applicable to mass production of standardized items, such as bricks, concrete blocks, or spark plugs (which is already highly automated), nor to the production of ultra-high quality porcelains. For production of batches between these extremes, flexible manufacturing systems should be feasible. Robots would be appropriate to palletize extruded raw materials, load roller machines, unload formed workpieces, and molds to handle "capsules." The latter

involves unloading formed workpieces, dipping or spraying with glaze, grinding after the combustion process, and loading or unloading kiln carriers. This last requires complex but precise motions, since workpieces must be piled up in several stories on each carrier before going to the kiln.

- *Glass industry.* Already highly automated except for transferring workpieces to production machines from loop conveyers. To some extent, this can be done by existing robots. But robots with pattern recognition abilities will be much more useful since workpieces can become disordered by automatic unloading devices. Compliant grippers with tactile feedback will also have to be developed to pick up glass without breaking it.

- *Clothing Industry.* Robots should eventually be applicable to loading and unloading sewing machines. However, the problem of separating one piece of cloth from a pile, and of orienting it properly for sewing, remains incompletely solved. Labor costs in this industry are comparatively low, so automation is appropriate only for fairly simple tasks (i.e., subassembly operations) with long runs. Robots can be used in packaging and materials handling.

- *Woodworking industry.* Robots cannot be expected to handle logs, and sawmills are already highly automated because lumber products are quite standardized. Loading of machines for finishing operations would require specialized robots capable of long rectilinear motions. Palletizing is appropriate for existing robots, as are painting and varnishing operations (except possibly where very high-quality finish is needed). Robots will be appropriate for cutting, forming, and assembly operations, for example, in the furniture industry, when specialized grippers and tools are sufficiently developed.

FUTURE TECHNOLOGICAL DEVELOPMENTS

Areas of likely technological improvements of robots in the coming decade can be divided into three headings:

1. Structural design, actuators, and grippers

2. Controls (including design of algorithms)
3. Sensors

With regard to structure, it has already been remarked that the polar articulated architecture theoretically will improve, primarily because of compensation for systematic (i.e., predictable) components of mechanical variance, such as wear and variable payload. Also, statistical compensation for random inaccuracy will be incorporated in feedback control systems. Other changes to be expected include universal use of 6(+) axis systems, improved clamping capability through flexible joints or fingers, and proximity feedback controls (de Gregorio 1980).

Regarding controls, it can be expected that (digital) programmable computers will control nonservo point-to-point while minicomputers will be needed for servo-controlled and continuous path robots. The declining cost of electronics increasing computational power available, opens the possibility of using general-purpose hardware with specialized software for different applications. Software development is already a major expense to the robot manufacturers, and the availability of highly developed proprietary software adds significantly to the capabilities of hardware in some cases.[15] An advantage of this approach is that a unit in place can later be upgraded by means of improved software, thus reducing the threat of rapid obsolescence in a fast changing field.

Interest in developing off-line programming capability is directly related to the problem of integrating robots and machine tools together into computer-controlled manufacturing "cells." The factory of the future will feature a hierarchy of levels of control. At the lowest level, simple computers will become essential components of robots and machine tools. At higher levels, more complex computer networks will control and monitor workflow scheduling, machine availability, inventory, and the like (Nof and Solberg (1979); Fox, (1981b); General Motors (1982). The supervisory computers will receive information in real time from, and pass commands to, subsidiary minicomputer control cells of machines.

---

[15] This appears to be particularly true for Cincinnati Milacron, which has many years of prior experience developing software for NC machine tools. ASEA, another high-priced manufacturer, has concentrated on developing adaptive software (capable, for instance, of following and mimicking a three-dimensional contoured surface).

Similarly, the computer controlling a cell as a whole will receive data from, and pass commands to, the microprocessors controlling each individual machine. Again, the cell controller will be concerned with internal workflow, scheduling, machine availability, inventory (within the cell), and so on. The individual machine controller will handle setups for each operation to be done on that machine.

This hierarchical control system necessarily involves language interfaces at each level. For example, the CNC machines and robots must be controlled internally in explicit terms specifying tool or arm locations, speed, directions, and so on. Instructions from the operator (at present) or from the cell controller (in the future) will be couched in more general task-oriented terms, e.g., the locations and depth of holes to be drilled on work pieces and the locations of welds. At present, many robot manufacturers provide their own proprietary control languages, partly for competitive reasons. However, there is a growing sentiment (expressed, for instance, at the recent NBS/Air Force/CAM Workshop) that both explicit machine-oriented and implicit task-oriented languages should be standardized to permit interfaces between different machines and systems. Candidates for a standard explicit language include VAL, EMILY, SIGMA, and WAVE. Candidates for a standard implicit language interface include AL, ROBOTAPT, AUTO PASS, RAPT, and MAL.

The interface between controls and sensors has already been identified as a critical area for future development. The problem, in brief, is to design operating "meta"-systems software to permit real-time modification of current programs in response to incoming data from sensors. This is a brief characterization of the so-called feedback scheme that is needed to facilitate many robot applications.

The integration of microprocessors ("chips") with miniaturized signal detectors of various kinds will result in rapid cost reductions for digitized sensory data. This will be applied initially to the standard instruments, for example, thermometers, strain gauges, magnetometers, and optical scanners.[16] Cheaper signal

---

[16] Vision systems, not surprisingly, are getting the most attention. Solid-state optical imaging systems have been developed by Hughes Aircraft Co. ("Omneye'), Stanford Research Institute, Machine Intelligence Corp. (founded by a SRI researcher), Optek Corp.,

processing capability will ultimately permit the use of unconventional measuring devices such as laser interferometers and ultrasonic transducers.

In the longer term, it appears that basic research is most needed (though not necessarily adequately supported at present) in the following areas (de Gregorio 1980):

- Kinematic and dynamic analysis algorithms, with particular application to series-connected elements with many degrees of freedom
- Servo system and logical network analysis algorithms capable of dealing with multivariable, nonlinear structures
- Digital control through multiple microprocessor networks
- Optical recognition algorithms, perhaps "hard wired" into special-purpose chips
- Design of advanced flexible grippers (hands)
- Development of computerized factory management systems with provision for optimizing the use of robots within a larger context
- Reliability analysis for complex systems, involving robots, especially in difficult environments
- Development of "artificial intelligence" capabilities, meaning a software system in which the internal operating programs automatically self-adapt in real time, to changes in the external environment

The last item, in particular, is an active experimental pursuit but is regarded by industry experts as being outside the scope of commercial possibilities, at least for the 1980s.

REFERENCES

Abraham, R.G., T. Csakvary, and G. Boothroyd. *Programmable Assembly Research Technology Transfer to Industry, 5th Bi-monthly Report.* Technical Report NTIS #PB285886, Westinghouse R&D Center, 1977.

Automatix, Hitachi, and General Motors (licensed to Unimation) among others. Others developing vision systems include Hitachi, Kawasaki Heavy Industries, General Electric, Texas Instruments, IBM, and Renault.

Abraham, R.G., et al. *Programmable Assembly Research Technology Transfer to Industry, 4th Bi-Monthly Report.* Technical Report 77-6g1-APAAS-R4, Westinghouse R&D Center, 1977. NTIS #PB277978.

Agin, Gerald J. Computer Vision Systems for Industrial Inspection and Assembly. *IEEE Computer Magazine,* May 1980.

American Machinst. The 12th American Machinist Inventory of Metal-working Equipment, 1976–78. *American Machinist* 122(12):133–48, December 1978.

American Machinist Magazine. Machine Tool Technology. *American Machinist* 124(10):105–28, October 1980. An abstract of the Machine Tool Task Force Report, Technology of Machine Tools.

Asada, Haruhiki, and Takeo Kanade. *Design of Direct-Drive Mechanical Arms.* Technical Report CMU-R1-TR-81-1, Robotics Institute, Carnegie Mellon University, April 1981.

Ayres, Robert U. *Three Industrial Revolutions—Policy Implications.* Technical Report, Department of Engineering and Public Policy, Carnegie-Mellon University, June 1982.

Ayres, Robert U., and Steven M. Miller. Industrial Robots on the Line. *Technology Review* 85(4):35-45, May/June 1982.

Ayres, Robert U., Leonard Lynn, and Steven M. Miller. Technology Transfer in Robotics: US/Japan. In *Proceedings of the First U.S.–Japan Technological Exchange Symposium.* Japan-America Society, 1302 18th Street, N.W., Washington, D.C. 20036, 1981.

Barash, M.M. Computerized Systems in the Scheme of Things. In *1979 Fall Industrial Engineering Conference,* pages 239–46. American Institute of Industrial Engineers, 1979.

Birk, John R. and Robert B. Kelly. An Overview of the Basic Knowledge Needed to Advance the State of Knowledge in Robotics. *IEEE transactions on Systems, Man and Cybernetics* SMC-11(8):574–79, August 1981.

Bjorke, O. Computer Aided Parts Manufacturing. *Computers in Industry 1:3–9, 1979.*

Robert C. Bolles. *An Overview of Applications of Image Understanding to Industrial Automation.* Technical Report 242, SRI International, May 1981.

Carbonell, J., R. Mickalski, and R. Mitchell (editors). *Machine Learning.* Tioga Press, 1982.

Coiffet, P., et al. *Les Robots: Traite de Robotique.* Hermes, Paris, 1983.

Cook, Nathan H. Computer-Managed Parts Manufacturing. *Scientific American* 232(2):22–29, 1977.

Cowart, N.A., et al. *Programmable Assembly Research Technology Transfer to Industry - Phase 2.* Technical Report, Westinghouse R&D Center, 31 March 1981. Year-end report for NSF grant ISP 78–18773.

de Gregorio, G.M. Technological Forecasting on Industrial Robots. In *10th International Symposium on Industrial Robotss, Milan, Italy.* IFS Publications, Ltd., Bedford, England, 1980.

Dreyne, Darrel. *Unmanned Machining Centers.* Technical Report MS79-934, Society of Manufacturing Engineers, 1979.

Eaglen, Ralph L. Human Factors as Related to Improved Productivity. In Arthur R. Thompson, Working Group Chairman (editor), *Machine Tool Systems Management and Utilization,* pages 75–93. Lawrence Livermore National Laboratory, October 1980. Volumn 2 of the Machine Tool Task Force report on the Technology of Machine Tools.

Engelberger, Joseph F. *Robotics in Practice.* American Management Association, New York, 1980.

Fisher, E.L., and S.Y. Nof. *The Work Abilities of Industrial Robots: A Survey and Analysis.* Technical Report, School of Industrial Engineering, Computer Integrated Manufacturing Laboratory, Purdue University, March 1982.

Fox, Mark S. *Reasoning With Incomplete Knowledge in a Resource-Limited Environment: Integrating Reasoning and Knowledge Acquisition.* Technical Report CMU-RI-TR-81-3, Robotics Institute, Carnegie-Mellon University, March 1981.

Fox, Mark S. *The Intelligent Management System: An Overview.* Technical Report CMU-RI-TR-81-4, Carnegie-Mellon Robotics Institute, August 1981.

Froelich, L. Robots to the Rescue? *Datamation*:81–96, January 1981.

General Accounting Office. *Manufacturing Technology—A Changing Challenge to Improve Productivity.* Government Printing Office, Washington, D.C., 1976.

1982 General Motors Public Interest Report. Technical Report, General Motors, April 1982.

Gevarter, William. *An Overview of Artificial Intelligence and Robotics: Volume II-Robotics.* Technical Report NBSIR 82–2479, National Bureau of Standards, Industrial Systems Division, March 1982. Prepared for the National Aeronautics and Space Administration Headquarters.

Golden, M.D., et al. *ICAM Robotics System for Aerospace Batch Manufacturing, Task A.* Technical Report AFWAL-TR-80-4042, ICAM Program, Wright Patterson Air Force Base, Ohio. Prepared by General Dynamic Corporation under subcontract to the ICAM program.

Hawegawa, Yukio. New Developments in the Field of Industrial Robots. *International Journal of Production Research* 17(5):447–54, 1979.

Hill, John W., Jan Kramers, and David Nitzen. *Advanced Automation for Shipbuilding.* Technical Report SRI project 1221, SRI Instructional, Industrial Automation Division, November 1980. Prepared for the Naval Material Command.

Holmes, John G. Integrating Robots into a Machining System. In *1979 Fall Industrial Engineering Conference*. American Institute of Industrial Engineers, 1979.

Jablonowski, J. Aiming for Flexibility in Manufacturing Systems: Special Report 720. *American Machinist* 124(3):167–82, March 1980.

Jablonowski, Joseph. Robots: Looking Over the Specifications. *American Machinist* 126(5)163–78, May 1982. Special Report 745.

The Specification of Industrial Robots in Japan. Japan Industrial Robot Association, 1982.

Understanding Manufacturing systems. Kearney and Trecker Corporation, Milwaukee, Wisconson, 1978.

Kelly, Robert B, John R. Birk, Henrique Martines, and Richard Tella. A Robot System Which Acquires Cylindrical Workpieces from Bins. *IEEE Transactions on systems, Man, and Cybernetics* 12(2):204–13. March/April 1982.

Kinnucan, Paul. How Smart Robots Are Becoming Smarter. *High Technology*:32-40, September/October 1981.

Lerner, Eric J. Computer Aided Manufacturing. *IEEE Spectrum* 18(11):34–39, November 1981.

Machover, Carl, and Robert Blough. *The CAD/CAM Handbook*. Computervision Corporation, 1980.

Pam McCordick. *Machines Who Think*. W.H. Freeman Press, San Francisco, California, 1979.

Merchant, M. Eugene. *The Factory of the Future—Technological Aspects*. Technical Report, Cincinnati Milacorn Inc., Research Planning Department, 1980. Presented at the 101st Winter annual Meeting of the American Society of Mechanical Engineers.

Moravec, Hans P. *The Endless Frontier and the Thinking Machine*. Technical Report, Stanford Artificial Intelligence Lab (SAIL), 1978.

Moravec, Hans. P. *Obstacle Avoidance and Navigation in the Real World by a Seeing Robot Rover*. PhD thesis, Stanford University, 1981. Reprinted by Carnegie-Mellon University Robotics Institute, 1981.

Mechanical Production Research Program, NSF *Ninth Conference on Production Research and Technology*, National Science Foundation, Washington, D.C., 1982.

Nof, Shimon Y., and James J. Solberg. *Control Considerations in Layout Configurations for Manufacturing Systems*. Technical Report, American Society of Mechanical Engineers, 1979.

Nof, Shimon Y., Moshe Barash, and James J. Solberg. Operational Control of Item Flow in Versatile Manufacturing Systems. *International Journal of Production Research 17(5):459–89, 1979*.

Paul, Richard P., and Shimon Y. Nof. Work Methods Measurement—A Comparison Between Robot and Human Task Performance. *International Journal of Production Research* 17(3):277–303, 1979.

Raibert, Marc. H. Dynamic Stability and Resonance in a One-Legged Hopping Machine. In *Symposium on Theory and Practice of Robots and Manipulators*. Warsaw, Poland, 1981.

Raibert, Marc H. *Robotics in Principle and Practice*. Technical Report, Robotics Institute, Carnegie Mellon University, Pittsburgh, Pa., 1981.

Ralston, Anthony (editor). *Encyclopedia of Computer Science*. Van Nostrand Reinhold Company, New York, 1976.

Reichardt, Jasia. *Robots: Fact, Faction and Prediction*. Penguin Books, 1978.

Robot Institute of America. *Robot Institute of America Worldwide Robotics Survey and Dirctory*. Society of Manufacturing Engineers, Dearborn, MI., 1982.

Rosenberg, Nathan. *Technology and American Economic Growth*. M.E. Sharpe Inc., New York, 1972.

Schraft, R.D., E. Schults, and P. Nicolaisen. Possibilities and Limits for the Application of Industrial Robots in New Field. In *10th International Symposium on Industrial Robots, Milan*. IFS Publications, Ltd., Bedford, England, 1980.

Sutton, George P. Training and Teaching. In Arthur R. Thompson, Working Group Chairman (editor), *Machine Tool Systems Management and Utilization,* pages 8.6-1 to 8.6-10. Lawrence Livermore National Laboratory, October 1980. Volumn 2 of the Machine Tool Task Force report on the Technology of Machine Tools.

William R. Tanner, editor. *Industrial Robots: Fundamentals*. Society of Manufacturing Engineers, Dearborn, MI., 1979.

Taramin, Khalil (editor). *CAD/CAM: Meeting Today's Productivity Challenge*. Society of Manufacturing Engineers, Dearborn, MI., 1980.

Toepperwein, L.L., and M.T. Blackman. *ICAM Robotics Application Guide*. Technical Report AFWAL-TR-80-4042, Volume II, ICAM Program, Wright Patterson Air Force Base, Ohio, 1980. Prepared by General Dynamics Corp. under subcontract to the ICAM program.

Weck, M., Y.K. Zenner, Y. Tuchelman, and D. Zuhlke. Concept of Integrated Data Processing in Computer Controlled Manufacturing System (FMS). In *Vth International Conference on Production Research,* Technical High School, Twente, the Netherlands, 1979.

Woodbury, Robert S. *Studies in the History of Machine Tools*. The MIT Press, Cambridge, Mass., 1972.

Wright, Paul K. and Alex Holzer. A Programmable Die for the Powder Metallurgy Process. In *Proceedings of the 10th North American*

Manufacturing Research Conference, pages 231–38, 1981. Held at Penn State University, State College, Penn.

Yoshikowa, Hiroyuhi Keith Rothmill, and Josef Hatuany. *Computer-Aided Manufacturing: An International Comparison.* Technical Report, National Academy Press, 1981. Sponsored by the National Research Council.

3 INTRODUCING ROBOTS AND FLEXIBLE PRODUCTION TECHNOLOGIES INTO MANUFACTURING

The questions addressed in this chapter are, in brief:

- *Why* are robots being used in manufacturing?
- *How* is the decision to implement usually arrived at?
- *What* adaptations are typically required in the user organization?

Because of the nature of the questions, it is evident that many of the answers can best be provided by users themselves. Hence we have made extensive use in this chapter of surveys and survey articles by academics and journalists. The major survey and/or interviewing of some forty major U.S. corporate users was carried out by a group of (mainly) graduate students at Carnegie-Mellon University (1981), with the cooperation of the Robot Institute of America (RIA). It will be referred to as the CMU survey. A copy of the survey questionnaire used and a summary of the responses are included in the Appendix. Interview respondents were promised anonymity, so corporations were given code letters (A, B, C, etc.) and identified only by industry. The last section in this chapter, *Organizational Adaptations*, draws on the above survey but is primarily based on a recent study of workers psychological reactions to robots and of the implementation process conducted by Professor Linda Argote, Professor Paul Goodman, and David Schkade of the Graduate School of Industrial Administration at CMU.

MOTIVATIONS FOR RETROFITTING ROBOTS IN EXISTING FACTORIES

The various motivations to install robots in manufacturing plants have been discussed at length. Primary among them is the desire to reduce labor costs. Other motivations are significant, however. They include:

- Improved quality
- Elimination of tedious and dangerous jobs
- Increased output
- Material savings
- Flexibility

Reduced Labor Cost

In his recent book, Joseph Engelberger (1980a: 103), president of Unimation Inc.—the largest U.S. robot manufacturer—asserts: "The prime issue in justifying a robot is labor displacement. Industrialists are mildly interested in shielding workers from hazardous working conditions, but the key motivator is the saving of labor cost by supplanting a human worker with a robot." The CMU survey and all other available evidence strongly supports this view, e.g., (Whitney et al., 1981; Industrial Robot 1981; Ciborra, Migliarese, and Romano 1980). Respondents to the survey overwhelmingly ranked reduced labor costs as their main motivation for installing robots. Several respondents in various metalworking industries (except autos) indicated that the "true" cost of an hourly paid employee to them (in 1981) exceeded $30,000 a year. Assuming 2,000 working hours to a year, this comes to over $15 per hour. For most auto industry production workers, hourly labor wage estimates, including all benefits and support costs, already reached as high as $20 in 1981. A recent Delphi survey of social scientists and industry managers by Arthur Andersen Company and the University of Michigan reported that *direct* hourly wage rates for less skilled workers in the automobile industry are expected to reach $18 per hour by 1990, double the 1979 level (Arthur Anderson & Company/The University of Michigan 1981). Wage rates for skilled workers were estimated to reach $25 an hour by 1990, a 125 percent increase beyond 1979. High and rising labor cost can be expected to accelerate the utilization of labor saving technology in general, and robotics in particular. General Motors Chairman Roger B. Smith was quoted as saying, "Every time the cost of labor goes up $1 an hour, 1,000 more robots become economical."[1]

Figure 3–1, an estimate of hourly robot-human cost differentials compiled by Engelberger, clearly shows how the robot manufacturers are "selling" the potential for labor cost savings. According to these figures, the "all included" cost of a medium priced robot in 1982 is $6.00 an hour, and hourly operating cost have only risen from 50 percent (from $4.00 to $6.00) since 1967, even accounting for the increase in interest rates. In comparison, hourly compensation for manufacturing workers, including fringe costs, has risen by nearly 180 percent during the same period (see Chapter 6). Engelberger comments that "if they ignore that difference, they ignore it at their peril."[2]

There are some questions, to be sure, as to whether Engelberger's estimate of the "all included" cost of a robot is as comprehensive as the "all included" cost for labor. He maintains it is, but users and consultants often point out that the cost of the robot only accounts for a fraction (sometimes one-third or less) of the total cost of designing and implementing a *new* application. One respondent, from a firm with extensive robot experience, pointed out that many of the "intangible" indirect costs associated with bringing a robot on line and maintaining it are often overlooked. Even allowing for some understatement of true robot costs, there is widespread agreement among robot users that these machines do effectively cut costs, in addition to sometimes providing other benefits. Richard Beecher (1981: 15), head

---

[1] John Holusha, "GM Shift: Outside Suppliers," New York Times, October 14, 1981, page D-1.

[2] Lee Holtz, "He Created Technology From Science Fiction," Pittsburgh Press, May 14, 1982, page A-2.

Figure 3–1.   A Comparison of Robot and Human Costs.

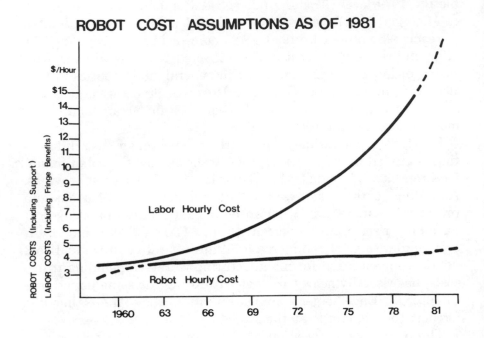

ROBOT COST ASSUMPTIONS AS OF 1981

ASSUMPTIONS

Unimate price	$50,000
Useful life	8 years at two shifts per day
Cost of money	15 percent
Installation cost	Two at $12,000 per installation
Maintenance	$1.15 per hour
Power cost	$0.40 per hour
Overhaul (Two)	$0.40 per hour
Depreciation	$1.56 per hour
Installation	$0.80 per hour
Money cost	$1.10 per hour
Hourly operating cost, 1981	$5.41

Source: Cost estimates supplied by Unimation, Inc.

of General Motors Robotics Applications Unit, states, "Based on our fourteen years of experience with the industrial robot, we know that it does, in fact, reduce operating costs, increase productivity, improve product quality, reduce scrap, reduce OSHA compliance costs, and improve our capital equipment." According to a recent survey of five German factories by Battelle Institute (Gizycki 1980), the average labor savings for robots installed to date in the five plants has been 1.5 workers per shift and 4.0 workers for all shifts.

Many of the respondents to the CMU survey suggested that the heavy emphasis on direct labor savings was imposed by the methods of financial analysis used to evaluate robot applications. Almost all the firms interviewed based their decisions to use robots on financial analysis of costs and benefits. One user pointed out that the return on investment (ROI) calculation will not be favorable unless there is a dramatic decrease in direct labor costs. He added that arguments about improving product quality or increasing production flexibility were nebulous and did not carry as much weight with the financial people as direct labor savings. Given that financial analysis requires short-term payback or high ROI, it is understandable that there is such a heavy emphasis on direct labor saving. Aside from increased output, it is the only variable that is quantifiable in a straight-forward fashion.

Elimination of Tedious and Dangerous Jobs

Many factory jobs range from merely dull to extremely tiring and unpleasant—even "dehumanizing." The following quotation taken from Studs Terkel's book *Working* (1974: 221-22) is from a spotwelder at a Ford plant describing his job:

> I stand in one spot, about two or three feet in area, all night. The only time a person stops is when the line stops. We do about 32 jobs per car, per unit. Forty-eight units an hour, eight hours a day. 32 × 48. Figure it out, that's how many times I push that button.
>
> The noise of it is tremendous. You open your mouth and you're liable to get a mouthful of sparks. (*Shows his arms.*) That's a burn, these are burns. You don't compete against the noise. ... It don't stop. It just goes and goes and goes. I bet there are men who have

lived and died out there, never seen the end of that line. And they never will—because it's endless. It's like a serpent. It's just a body, no tail. It can do things to you.

Perhaps not surprisingly, there have been worker rebellions, such as the General Motors Lordstown strike in 1972 (Business Week 1972). Dissatisfaction with such working conditions, especially on assembly lines, have been held partly responsible for labor unrest in the automotive industry. Indeed, the fact that assembly-line workers, in particular, perceive themselves as "slaves" to an inhuman machine—the moving transfer line— can be counted as a consequence not of automation *per se*, but of incomplete automation. In any case, jobs on the assembly line are so undesirable that the auto industry (in its profitable years prior to 1979) was forced to pay extraordinary wages to lure workers, while tolerating high absenteeism, frequent shutdowns of the line, and comparatively-low quality workmanship.

This is the background that enables robotics enthusiasts to speak plausibly of alleviating the necessity of utilizing human beings in "dirty, dangerous, and degrading inhuman tasks which expose them to heat, dust, noxious fumes, and require them to handle hot parts and position them accurately between the jaws of machines that could easily maim a human worker" (Murphy 1976).

There are some existing jobs (and some potential ones) that are not merely unpleasant but actually dangerous. One of the reasons for the Occupational Health and Safety Administration (OSHA), with strong support from organized labor, is the fact that 15,000 or so workers are killed on the job each year and 10 million are injured. Working with hot castings or forgings (discussed in section 2.5) is one example of a dangerous job. Handling radioactive materials, asbestos, or toxic chemicals would be others. Another example might arise, say, in a plant where some hazard is discovered (e.g., by OSHA) that has existed for some time but was not recognized as such. A specific, though hypothetical, example might be the danger of inhaling small particles of cotton over a long period of time in cotton-thread spinning or weaving mills. Conceivably, it would be cheaper to replace all machine operatives by robots—even at higher hourly cost—than risk heavy future damage awards to former employees with respiratory problems. Finally, robots may be able to function in places that would be dangerous or inaccessible to humans, such as underground coal mines, pipelines, deep gold mines, underwater, or space (Chapter 4).

Respondents to the CMU survey considered the elimination of tedious and dangerous jobs second in importance only to labor savings. It should be noted that for some respondents this motivation was related to direct labor savings. One respondent (see Q Corporation) said that there is no difficulty in incorporating this motive into an ROI framework, that is, the elimination of a dangerous or unpleasant job can also reduce labor costs for a firm. On the other hand, an executive from L Corporation commented that the financial people at his firm do not accept justification based only on the amount of eliminating potential hazards to workers because it is difficult to quantify. Two respondents noted that robots had reduced the number of production-related accidents where they were used.

A respondent from a large multiplant firm speculated that this rationale is an important factor in early adoption to help gain the acceptance of robots by labor but would be less important later. More broadly, there would seem to be an obvious public relations value to this justification for the use of robots.

Increased output rate

The ability of robots to increase output is a benefit often stressed in the current literature, though planners are frequently unable to determine beforehand the extent to which it will be realized. Output increase of over 100 percent for specific applications have frequently been mentioned in the trade literature. General Motors, for example, is reported to be making use of 150 robots in its Lordstown, Ohio, assembly plant. State-of-the art technology helped make it feasible to increase hourly production by 20 percent while making use of 10 percent fewer workers (New York Times 1980). General Dynamics claims a fourfold output increase in drilling and routing of sheet metal used in wing panels. Two respondents using robots for arc welding reported a 50 percent increase in parts welded per hour. Another user claimed that as a rule of thumb, a robot used in machine loading or material handling applications will result in a 10 to 12 percent increase in throughput over a human operator since it works consistently without taking breaks. A respondent from Corporation F said, "Sometimes the productivity increase is so high we can justify the robot without removing direct labor." The potential for increasing output and the implications of such increases for manufacturing costs and for employment are discussed in more detail in Chapter 6.

Over two-thirds of the users contacted in the CMU survey claimed that the average length of the work shift per day increased when robots were used, which in itself would lead to an increase in output. Two users said that the average number of shifts actually decreased as a result of using robots, and all others said that there was no change in the number of shifts.

It must be recognized that increased output and decreased costs are two sides of the same coin. In most plants, overall input is controlled by a number of factors, and increased capability at one point, for example, on a production line, could not be utilized. Thus, where robots are retrofitted into existing plants (as is the case for most installations), increased output is not to be expected.

Increased Product Quality

Product quality was the second most important motivating factor for prospective users of robots and fourth for current users, based on the CMU survey. This may be because the search for improved product quality has intensified in recent years because of the increasingly high quality of imported foreign (especially Japanese) products. In the auto industry, this search for improved product quality has been a major motivating factor in the decision to invest in robotics. Chrysler Corporation, for example, claims that an effort to boost product quality was one of its major reasons for installing robotics in two of its "K-car" plants. Ninety-eight percent of the welding done at Chrysler's Newark, Delaware, plant is done by robots. It is hoped that more consistent high quality welds will cut warranty and repair costs as well as promote the sales of new cars. Similar hopes of improved product quality prompted General Motors Corporation to install two robotic welding lines in its Detroit Fleetwood body plant at a cost of $8.5 million. The robotic lines replaced only six welders, resulting in a labor cost saving estimated at only about $120,000 per year. Leslie Richards, plant manager at the time of the installation, views the robots rather as "an investment in quality" that has a ripple effect in cutting costs (Wall Street Journal 1981). According to Richards, "you have to overweld by 10 percent if it's done manually."

Among respondents to the CMU survey, all of those who ranked product quality as one of their three most important motivating factors reported that robots had indeed improved quality. Interestingly, more than half of the firms that gave quality less importance as a motivating factor also reported an improvement in quality. Almost two-thirds of users said that the percentage of parts rejectedin the robotic operation had decreased. The remainder said that there was no change in rejection rates. Four respondents said that quality control specifications had been tightened where the robot was used. Only one respondent reported a change in the number of quality control employees needed, and this was an increase. A small number of respondents acknowledged efforts to place a dollar value on product quality improvements, but most noted that this is a recent development.

Material Savings

Material savings are significant in a few robot applications. One application that has resulted in considerable material savings is robotic spray painting. At General Electric's Hotpoint dishwasher factory in Milwaukee, a single painting robot saved $19,000 worth of paint in one year while achieving improved coating uniformity (Engelberger 1980b). While some of this material saving occurred because of the increased spraying accuracy of the robot, a large saving resulted because the vacuum units installed to protect workers from paint fumes no longer had to be operated.

In automobile manufacturing, major savings can also be achieved in spray painting. According to General Motors vice-president, A.C. Mair, in a manual spraying operation only about 30 percent of the paint consumed by the plant gets onto the cars.

The remainder is deposited elsewhere or sucked into the giant exhaust systems that protect workers from fumes. General Motors hopes to improve spray efficiency to 50 percent by carefully "training" its (Trallfa) robot spray painters (Wall Street Journal 1981). Needless to say, the spraying robot is also unconcerned about the ambient air.

DECISION CRITERIA TO IMPLEMENT ROBOTS

Quantitative Evaluation

The decision to use robots at a firm typically begins with an initial proposal to install and is followed by a financial analysis and an engineering analysis of the application. When evaluating capital expenditure, the management of any profit-oriented firm is concerned with financial performance. Expenditures for robotic equipment are no exception. Robots may offer a wide variety of benefits, but if they cannot pass the test of a critical financial appraisal, it is unlikely that implementation will occur. One way to assess the worth of a robot is by means of an "equivalent wage" calculation, such as the one illustrated in Figure 3–1. However, it should not be assumed that a robot will (or should) replace a human worker the moment its equivalent wage is lower, since firms do not have unlimited access to capital and must always choose between competing alternative projects. Current industry practices for evaluating the desirability of a particular robotic (or other) capital expenditure include either a calculation of the payback period, a calculation of the rate of return on investment, or a discounted cash flow analysis. The payback period calculation estimates the length of time it will take to recover investment costs. In a simple case, assuming a robot replaces one man (per shift) and that throughput is unchanged by the substitution of a robot. The standard payback formula for a simple case is as follows (Engelberger 1980b):

$$P = \frac{I}{L - E}$$

where

P = the payback period in years

I = total capital investment required in robot and accessories

L = annual labor costs replaced by the robot (wage and benefit costs per man times the number of shifts per day)

E = annual maintenance costs for the robot

In a more generalized situation, the robot may work faster (or slower) than the human, in which case the output of associated equipment, such as machine tools, increases or decreases. Assuming the fractional speedup (or slowdown) factor is q, we get

$$P = \frac{I}{L - E \pm q(L+Z)}$$

where Z is a measure of the annualized value of the associated equipment. A common (but not necessarily theoretically ideal) measure of Z might be the annual depreciation. It is clear on reflection that other contributions, such as reduced materials use, may also contribute to the projected savings (the denominator).

The ICAM Robotics Application Guide (Toepperwein and Blackman 1980) claims that most robots can be expected to reach payback in one to four years. In plastics molding parts removal applications, for instance, it has been estimated that the payback through reduced labor alone ranges from three to sixteen months (Jones 1978), which assumes a 25 percent tax/benefit package for a single machine operator on a three shifts per day, five-day week schedule.

Another method of analysis that is more appropriate for a longer-term comparison of alternative investments is to calculate annual return on investment (ROI).[3]

$$ROI = 100 \times \frac{S}{I}$$

where

S = annual savings resulting from the robot = $L - C \pm q(L + Z)$

L = annual labor costs replaced by the robot

C = annual robot cost = $I/N + E$

---

[3] See Hill, Kramerm, and Nitzen (1980), Toepperwein and Blackman (1980), Whitney, et al., (1981), and Fleischer (1982) for more extensive discussions of the cost and benefits of robot use.

I = total investment in robot and accessories
E = annual maintenance costs of the robot
N = useful life of the robot, in years.

Another approach that is used when the costs and savings are not uniform from year to year is the benefit-cost ratio derived from a discounted cash flow analysis. In this case, the numerator is the sum of projected benefits over the life of the system and the denominator is the sum of projected costs, both discounted to their present worth. This third approach is used where major capital investments with a long lifetime are being evaluated. Over several years or more it is quite unrealistic to assume constant yearly savings, even in "constant" dollars. Labor costs over the past few decades have increased significantly faster than inflation, as shown in Figure 3–1. Energy costs since 1974 and interest costs since 1978 have also outrun other costs. On the other hand, computer costs have been dropping in absolute terms, and robot costs are likely to do the same as markets grow and manufacturing experience accumulates.

Ignoring, for the moment, distinctions within the robot "family," robot manufacturing cost C is likely to be governed by the "experience curve," which states that each time experience doubles, costs can be expected to decrease by a fraction a, where a is typically between 0.2 and 0.4. In mathematical form

$$C = cN^{-b}$$

where

$$b = \frac{-\ln(1 - a)}{\ln 2}$$

and N is cumulative production to date (experience).[4]

By definition, annual output Q is given by

$$Q = \frac{dN}{dt}$$

So, if Q were an exponentially increasing function of time, N would be a proportional exponential also. If

---

[4] For a good summary of learning curve theory and applications, see Nanda and Adler (1977).

$$Q = qe^{kt}$$

then

$$N = \frac{q}{k} e^{kt} = \frac{Q}{k}$$

and

$$C = c\left(\frac{q}{k} e^{kt}\right)^{-b} = c\left(\frac{q}{k}\right)^{-b} e^{-bkt}$$

Thus, for example, if $a = 0.3$ (whence $b = .515$) and $k = 0.35$ (i.e., the market grows at 35 percent per year—a seemingly modest projection for the next few years), then costs would drop at an annual rate of $0.515 \times 0.35 = 0.18$ or 18 percent. If the market grows at 50 percent per year, cost would drop nearly 25 percent per year!

This kind of price reduction, combined with rising labor costs, introduces a new kind of problem for the would-be robot purchaser, because the net discounted present value of the cost of a robot purchased in a given year over time, while the benefits increase over time as suggested by Figure 3–2. The meaning of these curves is that there is an incentive on the part of potential users to delay making major commitments. Indeed, this incentive to delay could conceivably interfere with the optimistic market growth forecast, which is part of the original basis for believing robot costs (and prices) will drop rapidly. The importance of this phenomenon is difficult to assess, though there is little doubt that it is real.

Both payback and return on investment analysis tend to focus on direct labor costs and other direct costs because these are easily quantifiable. They do not readily permit inclusion of other benefits such as improved product quality or greater manufacturing flexibility. More sophisticated forms of analysis are necessary to deal with the latter issue.

Quantifiability of Other Robot Benefits

It is important to note that there may be relatively unquantifiable but extremely important factors in a decision to replace

Figure 3–2.   Discounted Present Worth of Costs and Benefits of a Robot.

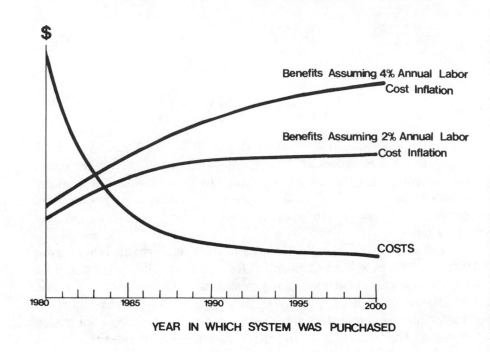

Benefits Assuming 4% Annual Labor Cost Inflation

Benefits Assuming 2% Annual Labor Cost Inflation

COSTS

1980   1985   1990   1995   2000

YEAR IN WHICH SYSTEM WAS PURCHASED

Source: Adapted from Katzman (1981).

human workers. One of these is exemplified by the current situation at H Corporation, a diversified manufacturer of capital and consumer goods. There appears to be a significant difference in the quality of labor-management relations in relatively small plants (less than 1,000 employees) compared to large plants (5,000 and up). For this reason, H Corporation management has decided to build all future plants for no more than 500 employees and to limit employment expansion in each plant to no more than 1,000. In some cases, this limit has essentially been reached, and increased output can be achieved only by increasing the productivity of existing machines and workers.

Robots can yield other benefits in addition to direct labor savings, including product quality improvements, increased throughput, better worker satisfaction, reduced indirect material requirements, increased reliability in production, and increased production flexibility. The CMU survey indicates, however, that most firms do not attempt to quantify "indirect" benefits in their financial analysis of the robot. Apparently, these types of costs are widely regarded as too difficult to quantify. Prospective users have difficulty calculating the value of indirect benefits, at least before they have a historical record of results. These records take several years to accumulate, even if accountants are systematically monitoring all the relevant variables. Direct labor savings, on the other hand, are relatively easy to quantify by using the direct wage rate and standard overhead factors.

The question was raised in interviews with corporate users of robots as to whether experienced users learn how to quantify "indirect benefits" as they accumulate experience with using robots. An executive at one firm speculated that inexperienced users only take direct labor cost into account because they do not know what other categories of cost will be affected. He said that after two years of robot use, his firm had learned how to quantify other indirect benefits, such as improved product quality and reductions in indirect material requirements.

Other experienced users did not report this kind of "learning." An aerospace components manufacturer indicated that his firm continued to justify its applications on the basis of labor savings alone. A large transportation equipment manufacturer said that after twenty years of robot use, his firm's motivation was still labor saving. A representative from another firm in the same industry commented that in six years of using robots, the firm had not changed its financial analysis (or motivations) significantly. Reducing direct labor costs was still the prime reason for using robots. Thus, there are differing opinions among experienced users as to whether the analysts learn how to quantify indirect benefits as they gain additional robot experience.

The executive from B Corporation speculated that transportation equipment manufacturers (his own firm included) would be spending billions of dollars in the next few years on new tooling (including robots) for reasons other than direct cost savings. He said that the main motives were to "improve their competitive position vis à vis the Japanese and other American manufacturers." At one or two other firms, managers reported investing heavily in robots because they wanted to "leapfrog" the competition. They thought they could do this by becoming adept with an emerging technology which potentially could give them superior manufacturing capabilities.

Thus, broader strategic concerns, such as long-term competitiveness, apparently are considered, yet they are seldom mentioned as the most important motivations. Only one firm, B Corporation, said outright that they had invested heavily into robotics to improve the quality and the competitive standing of their product. They were also the only firm to give strong emphasis to other intangibles, such as improved production flexibility. Interestingly enough, this spokesman was the only person among all those interviewed to say that robot applications were not evaluated primarily on the basis of ROI or payback period.

FLEXIBLE MANUFACTURING TECHNOLOGIES

There are two contexts in which there exists a clear need for greater flexibility with respect to resource allocation and planning. The first context occurs when a resource, such as a machine, is used for different tasks within a given time period. Job shops and batch production facilities use general purpose machines capable of producing a variety of products. The second context occurs when a resource is used for only one task within a time period but where there is uncertainty as to how long that task will be required. Thus, there may be a need to reorganize resources to generate new alternatives e.g., to respond to shifts in the market.[5] This is a potential problem for all manufacturers, but especially for high-volume producers who have always been forced to specialize their resources and organizational structure to a high degree to achieve economies of scale.

The current use of the term *flexible manufacturing* refers to an evolutionary development beyond both conventional batch manufacturing and mass production technology, in which the versatility of the former is combined with the high operating rates and low unit costs of the latter. Flexible manufacturing implies the use of programmable, computer controlled equipment both to operate the machines and to coordinate production lines or processes. The use of computers provides a key difference in terms of speed and controllability between the previous generation of manually controlled general purpose machines and the emerging generation of "programmable" automation. The use of *software-based*, high performance, general-purpose control systems provides a key difference between the previous generation of "hard-wired" mass production machines and the more general purpose "soft-wired" types of flexible manufacturing systems.

Industrial robots are a key component of flexible technologies because their reprogrammability allows them to be quickly adapted to changes in the production process. A robot may be programmed to perform the same task on a variety of different work pieces. This type of flexibility is commonly seen in several application areas, such as spot welding and materials handling. A second type of flexibility involves shifting the robot to an entirely new task. Despite the frequent mention of this factor in the trade literature, it was apparently not an important motivation to most of the respondents of the CMU survey. Evidence from interviews suggests that, after installation, robots have rarely been transferred to new applications. In I Corporation, a robot no longer needed for its assigned task, taken off line, sat idle because no one in the division wanted to invest the time to adapt it to a new application. Two respondents said that using robots had enabled them to manufacture more complex designs (another type of

---

[5] Professor Herbert Simon, at Carnegie-Mellon University, suggested these two "dimensions" of flexibility. For other discussions of flexibility, see Whitney et al. (1981), Table 5–10, and Warnecke, Bullinger, and Kolle (1981).

production flexibility), and eight claimed that robots had reduced setup times in the operations where they were used, which reduces the cost of maintaining flexibility.

There was a large spread in the priority given to flexibility by respondents. Some manufacturers, such as B Corporation, did view increased flexibility as an important impact. This view seems most prevalent in industries with frequent major model changes, such as the auto industry. However, at present, robots are not widely viewed by most users of robots as an investment in flexibility.[6] This is almost certainly due in part to the fact that respondents were generally concerned with applications in existing plants, not new (or redesigned) plants. In addition, many engineers and planners do not have have experience with the flexible production technologies. A senior engineer from O Corporation complained that "production engineers are so used to thinking in terms of hard automation that it is difficult to get them to think in terms of using programmable automation."

Nevertheless, it is widely acknowledged in engineering circles that for batch production, flexible automation—or flexible computerized manufacturing systems (FCMS)—are the wave of the future. Of course, it is important to acknowledge that this forthcoming change involves technologies other than robots. For this reason it does not seem appropriate to regard this restructuring of manufacturing as if it were an "impact" of robotics. On the contrary, we view the development of FCMS as of importance in itself and an offshoot of group technology and CAD/CAM developments. The physical integration of computerized manufacturing cells into FCMS was discussed in Chapter 2.

"Group technology" is a way of putting together machines and operations to simulate assembly line techniques in batch production. It consists of working out a formal classification or taxonomy of parts and then rearranging the shop floor so that machines are grouped according to part families, and the set of operations needed for their production (Opitz and Wiendahl 1971; Allen 1980).

Flexible computerized manufacturing systems involve linking a series of manufacturing cells (such as those arranged in group

---

[6] There are some notable exceptions. A recent article in the *Wall Street Journal* (1981) reports that General Motors intends to manufacture "modular" engines (from three to six cylinders) using common equipment.

technology) by means of a very sophisticated control and transport system so that workpieces with a given range of variability can be automatically subjected to differing production processes as necessary. The development of minicomputers for control made FCMS practical. An FCMS consists of machine tools, robots, conveyors, gauging equipment, computers, and software, integrated as illustrated in Figure 2–7. Robots interact with numerically controlled machine tools and other equipment, controlling the sequence of operations. It is more integrated and more automated than a traditional "job shop" consisting of machines in isolation, operated by individual humans.

Mass production, at the other extreme, is characterized by highly specialized equipment and long production runs. The market has either stabilized at a high level or is almost assured of steady growth." Examples of mass-produced items are standard types of hardware, electrical fixtures, appliances, televisions, and, of course, automobiles. Taylor (1979: 3-76) describes the important characteristics of thespecialized "hard automated" transfer line used to produce automobile engines.

> Of importance to note is that the system is based on a large volume of repetitive but complex machining operations. Because of precision tolerance requirements in addition to volume production, large manufacturing capital costs are involved. Except over a very limited range, little flexibility is inherent in the system to accommodate change. Only a single product is made with very limited or minor variations, but under a manufacturing environment that is engineered to turn out the product in large quantities at minimum cost.

The last sentence reveals the inherent limitations of "hard automation" technology. It is the cheapest method of production precisely because each element in the system is dedicated to a single function, for which it is optimized. However, the major disadvantage of "hard-automated" mass production technology is vulnerability to changes in the marketplace. These types of changes may be forced by government regulation (e.g. auto emission controls) or by shortages or cartel pricing of essential resources (e.g., OPEC oil prices). The resulting induced demand for a different product (e.g., small cars with four-cylinder engines and front-wheel drive) can make hard-automated production facilities such as transmission plants obsolete long before they are fully amortized. In addition, changes may be mandated by the availability of improved designs or better materials (e.g., radial tires versus bias-ply tires).

For example, an automobile engine plant is virtually a single specialized machine capable of producing only a single product model. If the product is no longer salable, for any reason, the plant must be written off.[7] This explains why such immense amounts of capital—$150 billion in fifteen years—will be required to increase the average fuel economy of U.S. cars to levels expected in 1990. It also suggest why, in an increasingly uncertain and volatile world-wide market, a more flexible manufacturing technology would offer possible economic advantages to mass producers (Goldstein 1981). Thus, it makes sense to invest in mass-production technology only if the product itself is technologically mature and relatively unchanging or if the market is closed to outside competition.

Nowadays, either of the above assumptions can be proven unfounded.[8] The world has grown "smaller" and more interdependent, while the unchallenged technological and economic dominance that the United States enjoyed in the 1950s is only a memory. The world is less stable and less predictable than it once was. It is therefore less risky—hence, likely to be cheaper in the long run—for a manufacturer to invest in more flexible means of production, permitting rapid response to changing markets even at some sacrifice of efficiency.

The current tradeoffs between efficiency and flexibility in manufacturing are illustrated schematically in Figure 3–3. For a wide range of products and low levels of production for each product, a conventional "job shop" is required. At the opposite extreme, high-volume production of one or several standardized parts, the dedicated transfer line is the appropriate technology. To date, robots and CAD/CAM have play little role at either extreme. At present, their domain is mostly in between, where operations are

---

[7] All of the "Big Three" auto makers have had to convert eight-cylinder engine models to a six-cylinder or four-cylinder engine, and scrap virtually all of the existing equipment in the plant (Shackson 1980; Diponio 1982).

[8] The entire U.S. auto industry was caught unprepared by the rapid rise in petroleum prices initiated in 1974 by OPEC. As a consequence, large U.S. cars, especially V-8s, are being phased out. From 75 percent of General Motors' output in 1977, the V-8 will drop to 5 percent by model year 1983 (Agnew 1980).

Figure 3–3.   Efficiency versus Flexibility Tradeoffs.

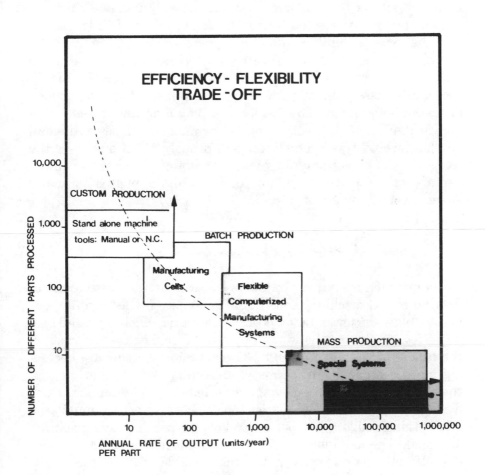

Source: Adapted from Queenen (1979).

repeated several hundred or thousands of times but where there is still a regular need for flexibility.

Where robots are installed as retrofits in existing conventional factories, their advantages cannot be fully exploited. Thus, we expect to eventually see a new generation of totally computerized factories not yet (or just fresh) off the drawing boards. This new kind of factory (suggested by the extension of the FCMS regime into the mass producer regime, in Figure 3–3) will be neither the traditional "job shop" consisting of a collection of stand-alone machines operated by skilled workers (or robots) nor the highly specialized mass production facility, as typified by an auto engine or transmission plant. The former will play a declining role, being increasingly limited to very complex or unusual shapes and small batches or prototypes. The latter will probably continue to be the technology of choice only for highly standardized parts or components that are produced in millions of units annually for many years without change of design.

Strategic Investment Issues

A key question for batch producers considering the possible introduction of NC machines and robots is: Where is the "indifference point" between lower labor cost and greater efficiency and productivity on the one hand, as opposed to higher capital cost on the other? While robotics and NC machining systems are becoming more competitive in terms of flexibility with human beings, their higher capital cost nevertheless reduces the manufacturer's flexibility in the financial dimension. The manufacture must seek and obtain enough new business to keep these expensive machines occupied. The key question for mass producers is slightly different. If production runs are long, flexible multipurpose NC machine tools may be more expensive to operate than specialized automatic machines that are optimized for specific tasks. A mass producer needs to identify the point of indifference between reduced efficiency and increased flexibility. Reduced efficiency implies higher unit costs. This could result from either more expensive capital, more expensive labor, or both. The choice of a production technology obviously depends on the manufacturer's assessment of environmental uncertainty. Higher front-end capital cost and/or operating costs for programmable automation could be justified, for instance, if the probability of an early write-off were sufficiently low. Or flexible, multipurpose equipment could be justified if it increased the average utilization rate of a hard-automated production line by substituting for specialized machines momentarily off-line for various reasons.[9]

The foregoing discussion, together with economic arguments given later in Chapter 6, provides some justification for our belief that three important phenomena will occur over the next several decades:

1. Flexible computerized manufacturing systems (FCMS) will become progressively cheaper to install, and more widespread.
2. Unit costs of batch production will steadily decrease as a result of using FCMS, reducing the differential in average processing cost per unit between batch and mass production.
3. The domain of FCMS (batch production in lot sizes of 100 to 10,000 units per year) will grow at the expense of other modes of production.

The extent of the expansion of FCMS into the existing domain of mass production is, of course, debatable. Mass production is often equated with the assembly line—where the concept started—but in actual fact, assembly lines are not particularly automated by current standards. (The auto assembly line is the most labor-intensive part of the industry.) Assembly lines in mass production sectors are obvious candidates for increasing future use of robots, especially after sensory feedback capabilities are further developed.

There is a more fundamental reason, however, why FCMS may encroach in the traditional mass production area. It is because of growing consumer demand for product diversification, or custom-

---

[9] While each machine in an integrated production line (e.g., for engines) may be 99 percent reliable, up-time for the line as a whole is unlikely to exceed 60 percent. This is because the whole line goes down when any machine in the sequence is down. The availability of flexible substitutes in such a situation could increase effective utilization of the entire facility. General Motors hopes to increase "up" time at its Windsor, Ontario, transmission plant by providing backup machine tools and alternative routings for workpieces to bypass off-line machines. To accomplish this, the mileage of automated materials handling gear in the plant is being increased tenfold (from two to twenty-three).

izing. Automobile manufacturers have, for several decades, successfully compromised by customizing the visible exterior and interior appointments (paint, trim, upholstery, optional accessories), while standardizing the rest. Thus, they get only the benefits of true mass production in the manufacture of standardized components. The final product is customized at the assembly stage by putting together standard parts in a unique combination. Nevertheless, there are limits to what can be achieved this way. It is no accident that most cars in a given size category have the same "look." And one cannot assemble a small car with front-wheel drive from parts for large cars with rear-wheel drive. To achieve true diversity of products, a more flexible manufacturing technology will be needed at the component level. Production runs will be shorter and changeovers more frequent. Most important, the need for extensive retooling to accommodate production redesign must be reduced or eliminated.

Curiously enough, the way to increase flexibility in the mass production (consumer goods) sectors seems to be to increase standardization of capital goods! That is, instead of expensive customized production lines optimized to a single product (or a small number of variants), there will be more and more cases where manufacturers will choose to build more flexible production facilities utilizing standardized programmable machine tools and robots. Specialized knowledge and specialization of the environment required to perform efficiently will be embodied in the software of the control systems, rather than in the mechanical design of the hardware.

An interesting implication of this trend is that an important existing inhibition on technological change in mass production industries will be relaxed. It is undeniable that an industry with heavy investments in mass production facilities will (indeed, must) resist introducing technological changes that would tend to make their existing products obsolete. This is because mass production technology is inherently inflexible—it cannot be adapted; it can only be scrapped and replaced. Rapidly evolving products tend to be batch produced. Mass-production, by its nature, is compatible only with stabilized or standardized products. Once mass-production facilities are in place, their very existence is a deterrent to innovation. Abernathy (1981: 4) has made this point with respect to the auto industry:

To achieve gains in productivity, there must be attendant losses in innovative capabilities, or conversely, the conditions needed for rapid innovative change are much different from those that support high levels of productive efficiency.

The unwillingness of the auto industry to seriously consider developing a replacement for the internal combustion engine (or, for that matter, the traditional steel frame and body) is entirely due to this fact of life. If the domain of the FCMS were to be extended into the auto industry life, a new era of automotive technological dynamism might become possible.

Robotics and advanced information systems can introduce flexibility into the manufacturing process, giving a firm the ability to inexpensively alter product designs and to efficiently produce a more diverse mix of products. This is in contrast with the current situation in which the organization is either specialized to the constraints of "efficient" but inflexible mass production or to the constraints of "flexible" but inefficient batch production. As the distinction between batch and mass production becomes blurred, and eventually eliminated with the development of high volume but flexible manufacturing systems, conventional types of organizational structures may no longer be appropriate. A challenge will be to devise organizational structures able to survive and to prosper in an international market with a greater diversity of products, higher quality products, and accelerated product life cycles.

ORGANIZATIONAL ADAPTATION

The introduction of the robot into the work place may have a significant impact on the worker, management, and the organizational unit. To effectively use this new technology, managers need to identify the critical points of impact and develop mechanisms to effectively adjust to this new technology.

Impact on the Worker—Displacement

An obvious effect of introducing robots into a factory is the displacement of workers. Displacement can mean losing one's job or being placed on another job. The CMU survey results indicate that over 70 percent of the workers whose jobs were displaced were retrained to either work with the robot or to perform other jobs within the plant. These results indicate that the initial type of displacement is moving workers to new jobs in the plant rather than causing them to lose their jobs. Important issues in this type of displacement are whether the new jobs are similar to displaced workers' previous positions and whether the new jobs entail higher pay. A recent study by Argote, Goodman, and Schkade (1982) indicates workers *are* very concerned about being "bumped."[10] That is, workers are concerned about being placed in a different, less desirable job with lower pay, or on a less desirable shift. Any of these types of displacement (job loss, less desirable job or shift, lower pay) were deemed as unsatisfactory by the workers.

At the present time it is difficult to assess the impact of displacement and its effect on worker resistance to change, since, with few exceptions, large numbers of robots have not been introduced into any work place. It was emphasized by most respondents to the CMU survey that no one has been laid off "because of a robot," at least in major corporations. Several of the larger firms interviewed stated that they had adopted a deliberate policy of not laying off workers as a result of introducing new technology. One firm pointed out that technologically induced layoffs would result in problems with the union.

Some interviews indicated that the number of jobs actually increased due to the necessary installation, maintenance, and servicing duties involved with the robots, especially when the machines were used to perform a new function not previously performed in the plant. Also, natural attrition to some extent mitigates the effects of substituting robots for workers. In the future, though, we expect that the rate of introduction of robots will almost certainly increase. If this leads to more incidences of job loss or bumping to less desirable jobs, then it would be realistic to expect greater negative work attitudes and greater resistance to change.

---

[10] The results of the research by Argote, Goodman, and Schkade (1982) will be referred to as the AGS study.

Impact on the Worker—The Robot Operator

The introduction of the robot in the factory will create some new job opportunities. Some production workers will have responsibility for the operation of the robot. One critical question is how the workers' skills and preferences fit the task requirements of the new job. Lack of fit between the worker skills and preferences and the job requirements can lead to negative worker attitudes and probably poorer performance; a good fit leads to worker satisfaction and better performance.

The CMU survey indicated that 60 percent of the firms surveyed indicated that the operator's job required greater skill levels. In response to this change in skill level, most firms in the survey began extensive retraining activities. Some organizations (68 percent) had help from the robot venders in training. Others developed their own in-house capability.

The AGS study indicates that retraining in the technical aspects of the job may not be enough. There may be some broader changes in job activities that are not susceptible to training. For example, in their study they reported that the new robot operator's job required more *watching* activities than *doing* activities. Operators may learn the *skills* of observation but still prefer doing activities (e.g., handling stock) that were taken over by the robot. This incongruency between the job activities and the preferences of the worker can generate feelings both of boredom and stress.

The introduction of the robot also can change worker interaction patterns on the job. The literature on automation indicates that if a new technology decreases workers' opportunities to interact with coworkers, one may expect increases in worker alienation. An important question is whether the introduction of a robot will substantially change formal and informal interaction patterns. The AGS study indicates that the introduction of a robot may lead to a reduction in opportunities for operators to interact with their coworkers.

A basic point in this analysis is that retraining can facilitate the adjustment of the robot operator. However, this new technology may, in addition, create a new set of activities that are not preferred by the worker or change interaction patterns. These changes may lead to an incongruency between the individual and the new job.

Impact on the Worker—Support Personnel

The introduction of the robot into the factory will have impact on a variety of support jobs such as maintenance, engineering, scheduling, and quality control. The analysis for the support worker is the same as for the operator. The robot can change activities and interactions. The CMU survey indicated that the robot increased skill requirements for the support jobs. Typically, experienced maintenance people required a week of formal training to adapt their skills to robots. One firm indicated that six to eight months of on-the-job training followed the schooling provided by the robot vendors. One auto assembly plant had recently hired a full-time educator to operate an ongoing in-house vocational college for the skilled trade workers (electricians, mechanical repair, etc.). The director of this program said that his company had no alternative but to insure that their workforce had the skills to maintain robots and other types of programmable equipment. At another plant, the AGS study indicated there was some frustration on the part of support personnel because they had to learn new skills and the resources for acquiring those new skills were not adequate.

In terms of interaction patterns, the robot is likely to increase the rate that support people interact with the robot operator and/or the supervisor in the department with the robot. The AGS study found the increased interactions were primarily of a cooperative nature. This was the first robot installation in the factory they studied, and both parties (operator and support) were highly motivated to get the robot up and running. A critical factor that will determine the future nature of the operator and support worker's relationship will be the number of robots in place and the resources of the support personnel to meet competing demands from the line personnel.

Impact on the Organization

While questions of organizational structure are often subordinate to questions of strategy and human resources, the introduction of the robot in the factory will have direct consequences for the organization itself. How the organizational structure is modified to meet this new technology will determine in part the effectiveness of the robot installation. The following are some of the possible points of impact on the organization:

1. Most robot installations will require modification of jobs which in turn will lead to job reevaluation and pay reclassification. Certain aspects of a job may be eliminated (e.g., material handling), while other activities (monitoring) are added. A way of evaluation of the skill requriements of the new job is needed. The AGS study found differences between management and worker expectations about classification of thejob of robot operator.
2. The robot may require new coordination mechanisms between support and line functions. As the robot requires closer coordination among line, maintenance, engineering, scheduling, programming, and related functions, there may be a need to develop formal coordination procedures.
3. There are obvious needs for greater retraining activities. Workers required to operate and support robots must have the necessary skills. In addition, workers who must move to other types work (whether inside or outside the factory) must be able to make the transition. The quality of the robot-related training will contribute to the effectiveness of the installation as well as to minimizing safety problems.
4. New safety procedures also may be needed. As with any machine, possibilities of accidents are present. Seventy-one percent of the firms surveyed by CMU indicated they had adopted some new safety measures to protect workers. The effects of the robots on accidents are not yet known. To the extent that workers are freed from dangerous or heavy manual activities, the chances for accidents are decreased. On the other hand, the AGS study found that some workers feel that robots can increase the potential for accidents.
5. New organizational units have appeared in some organizations for evaluating new technological advancements, planning for future applications, promoting applications, and facilitating change at given plants.

Introducing Robots into the Workplace

The above discussion outlines a variety of places the robot can affect the worker or organization. The nature of this impact can determine whether or not the robot will be an effective addition to the work place. One important issue concerns the process by which change is introduced. This can determine to some extent whether the effect is positive or negative. The process of introducing robots is probably more complicated than introducing other new technology. Because of the glamorous and farfetched portrayal of "super intelligent" machines in movies such as "Star Wars" and on television, there is a tendency to anthropomorphize robots. As a result, both workers and managers are more sensitive to their introduction into the work place than they are to the introduction of other technologies.[11] A variety of factors, both in terms of methods and procedures, affects the introduction of robots. Demonstrations of their use and limitations seems to be a powerful technique that workers feel aids in their understanding of the machine's use and limitations. The role of the first line supervisor is also very important. The more knowledgeable the supervisor, the easier the process of introduction. Some degree of worker involvement in introducing the new technology is also likely to ease the implementation process.

I. APPENDIX

Industry Policy Survey

The following survey is to be used in a study of the social impacts of robotics currently being done at Carnegie-Mellon University. All answers to this questionnaire are confidential and will be used only to compile a statistical study of industry's policies towards robotics. No specific respondents will be mentioned in the final report. Any questions about the survey may be addressed to Dr. Leonard Lynn at (412)578-2833. Those respondents who are not presently using robotics are directed to those questions marked by an asterisk. The following definition of ROBOT is suggested:

---

[11] See Argote, Goodman, and Schkade (1982:8).

"A Robot is a reprogrammable multi-functional manipulator designed to move material, parts, tools, and specialized devices through variable programmed motions for the performance of a variety of tasks" (adopted by Robot Institute of America, 1979).

QUESTIONNAIRE

*1. Industrial classification of establishment(4 digit S.I.C. if known).

*2. Total number of direct production workers in plant.

*3. Percentage breakdown of total production by batch size:

11	_____	custom (1-100 units/year)
10	_____	small batch (101-1,000)
4	_____	large batch (1,000-10,000)
11	_____	mass (over 10,000)

THESE ARE RESULTS FOR INDICATIONS OF GREATER THAN OR EQUAL TO 50 PERCENT.

4. In what year did you first introduce robots into your plant?

5. How many robots were in use in Jan. 1976? 121 Jan. 1981? 579

6. How many robots are now being used in each of the following uses:

159	_____	spot welding
72	_____	spray painting
190	_____	pick-up and place
14	_____	assembly
144	_____	other (specify) _____

7. Location of plant(s) where robots are used. State _____

*8. Please rank the following in order of importance as motivating factors for the use of robotics:

THE FOLLOWING ARE AVERAGES FOR THE RESPONSES, ZEROES ARE NOT COUNTED.

2.2   _____ to reduce labor costs.

5.9   _____ to ease compliance with OSHA regulations.

6.8   _____ to reduce capital costs.

3.7   _____ to improve product quality.

3.1   _____ to increase output rate.

5.1   _____ to increase the flexibility in changing products, designs, etc.

5.7   _____ to reduce material wastage.

3.0   _____ to relieve workers of tedious and/or dangerous jobs.

6.0   _____ to remedy high labor turnover problems.

9. With the introduction of robots, has work scheduling changed in any of the following areas?

   A. Average number of shifts per day:
      2   _____ decreased
      16   _____ no change
      0   _____ increased

   B. Average length of work shift per day:
      0   _____ decreased
      7   _____ no change
      11   _____ increased

10. Has the percentage of parts rejected changed?

    0   _____ increased rejection rate
    7   _____ no change
    11   _____ decrease in rejection rate

11. Have quality control specifications been tightened?

    4   _____ yes
    16   _____ no

12. Has the number of quality control employees changed?

    1   _____ increased
    18   _____ no change
    0   _____ decreased

13. Have robots had any of the following effects on production capabilities? Please check if yes.

    0   _____ ability to manufacture more complex designs
    2   _____ increased efficiency in raw materials usage
    6   _____ reduction in set-up times
    2   _____ both 1 and 3

14. Were robots placed in a new plant or in an existing plant?

    2   _____ new plant
    15   _____ existing plant
    3   _____ both

*15. Do you plan to place new robots in new plants or in existing plants during the next five years?

    0   _____ new plant
    20   _____ existing plant
    14   _____ both

16. How were operatives made aware that robots would be introduced into their plant?

    0   _____ through labor union
    14   _____ by meeting with management
    2   _____ workers were not notified prior to the introduction of robot
    2   _____ other (please specify)
    2   _____ combination of two of the above

17. How far in advance was the workforce notified (in months)?
    _____ 4.35

18. Which method was used most often to fill newly created positions?

    18   _____ retraining present workers
    2   _____ hiring new employees with required skills

19. If positions were eliminated, what percent of former workers were:

    3   _____ transferred without retraining
    10   _____ retrained for a new position within the plant

0   _____ retrained for a new position in another plant
0   _____ laid off
0   _____ retired early

THESE ARE FOR GREATER THAN OR EQUAL TO 50 PERCENT.

20. In any retraining efforts, did the unions:

1   _____ choose who was to be retrained
8   _____ support your intentions, but did not actually help in the retraining
2   _____ help to retrain the workers
0   _____ oppose your efforts

21. How have robots affected the level of skill required in the following areas?

A. Operatives:
   12   _____ increased skill level
   7   _____ no change
   1   _____ decreased skill level

B. Maintenance:
   18   _____ increased skill level
   2   _____ no change
   1   _____ decreased skill level

C. Production scheduling:
   2   _____ increased skill level
   17   _____ no change
   0   _____ decreased skill level

D. Materials handling:
   4   _____ increased skill level
   12   _____ no change
   3   _____ decreased skill level

E. Other (please specify) _____
   2   _____ increased skill level
   1   _____ no change
   0   _____ decreased skill level

22. Does your company have a counselor or psychologist to deal with worker problems caused by robots?

    3 _____ yes, we have always had a counselor
    1 _____ yes, we have had one since the introduction of robots
    0 _____ yes, we have had one for the past year
    1 _____ no, but we are considering it
    13 _____ no, it is unnecessary

23. To your knowledge have there been psychological problems caused by the introduction of robots?

    3 _____ yes
    17 _____ no

24. Since the first use of robots, the number of production-related accidents have:

    2 _____ decreased
    13 _____ no change
    0 _____ increased

25. Was it necessary to develop new worker safety measures?

    15 _____ yes
    6 _____ no

26. Did the supplier of the robot help with retraining?

    13 _____ yes
    6 _____ no

27. How long was the retraining period? _____ The following are averages

    Maintenance—25.3 DAYS
    Production Scheduling—21 DAYS
    Operatives—27.3 DAYS
    Other—20.8 DAYS

28. Have any new organizational units been created as a result of the introduction of robots or CAM?

6   _____ yes (please identify) _____
14   _____ no

29. Were any departments consolidated or eliminated?

0   _____ yes (please identify)
20   _____ no

30. How has the use of computer controlled production affected the number of managers required?

3   _____ decreased
18   _____ no change
0   _____ increased

31. Has an automated inventory control system been implemented?

10   _____ yes
8   _____ no

32. Have new robot-specific management positions been created?

5   _____ yes
17   _____ no

33. If yes, how were these positions staffed?

1   _____ retraining of present managers
1   _____ promotion of workers
1   _____ hiring of outsiders
1   _____ combination of 1 and 3
1   _____ combination of 2 and 3

INTERVIEW SUMMARIES

To supplement and expand on the results of the survey, a series of personal and over-the-phone interviews were conducted. Information was obtained from a total of thirty-one interviews with twenty-six corporations representing the various segments of industry in the United States. Included in the interviews were both firms who had responded to the written survey, and numerous major robot users who had not filled out the questionnaire.

In this section, summaries of a selected set of interviews are presented. Within each transcript, all references revealing either an individual or corporate identity have been replaced by a randomly selected letter of the alphabet to ensure confidentiality.

The objective of the interviews was to obtain supplemental information on areas such as:

1. Motivations for using robots
2. Decision to install robots and where to install them
3. Major positive and negative impacts of robots
4. Internal adaptations required to support robots

A CORPORATION, April 13, 1981

A corporation is a manufacturer of transportation equipment. They first started using robots in 1960, and now estimate that 370 are in use in their production facilities. The primary motivation for installing robots was to reduce total costs. The decision to invest in robots is based on the calculation of both return on investment and payback period.

The introduction of robotics has allowed the plant to increase the number of shifts. The use of robots has not led to a reduction in setup times; they are being used to manufacture the less complex, more routine designs. They have been placed in existing plants but on new product lines. For safety purposes, A Corporation has had to fence the robot areas off.

Workers displaced by robots are retrained within the plant; job openings due to attrition are sometimes not replaced. Some new personnel including mechanics have been hired from outside the plant.

One official indicated that the workers did experience some psychological problems. These problems were attributable to the fact that robots often broke down. A Corporation's general attitude toward robots is that they are unreliable and it is difficult to make them pay off.

A CORPORATION follow-up interview: April 24, 1981

The representative stated that the variables used in the return on investment (ROI) analysis are those used to evaluate robots.

Q. Is there a difference in the variables you quantify in the ROI analysis now as compared to when you started using robots?

A. Basically the same variables are in the ROI analysis. In other words, there has been no systematic attempt to extend the range of the benefits based on previous experience with robots. Some of the costs and benefits have changed though, because the robots have improved, and become more reliable. Maintenance costs have decreased.

The process of evaluating a robot for a retrofit is different than the process of evaluating the robot as one of several alternatives in an existing line. In the early applications, they were retrofitted. Later on, they were chosen as one of several alternatives when a new line was put together. In the retrofit situation, you already have a certain amount of hard automation present, and one does not replace it by adding a robot. When building a new line, it is a question of choosing among different types of automation.

B CORPORATION, April 20, 1981

B Corporation manufactures transportation equipment. Their motivations for implementing include four equally weighted factors: (1) to improve product quality, (2) increased flexibility in changing products and designs; (3) to eliminate the problems of high turnover rates in tedious and undesirable jobs; and (4) to remedy problems of severe labor shortages.

The idea to implement robots originates within the group operations level which is between the corporate and the plant level. This is a production engineering group which is active in the processing and tooling. It was said that this group is characteristic of all of the major firms in the industry. The group operations staff also picks the application areas, makes the final installation decision, evaluates the robot's performance, and does planning for future applications.

The decision to install robots involves cost studies and payback period determination. They estimate a robot's cost at $100,000 for purchase and installation. It costs about $30,000 a year including benefits to employ a worker. So a robot working two shifts will save the firm $60,000 with a payback period of about two years. In only one case was a robot taken off-line, and this was a test robot. In general, B Corporation has a positive attitude toward robots and reports no problems other than not having enough money to buy more.

The use of robots in the plant has not affected the shifts, and no safety or psychological problems have been encountered, however, the robots are fenced off from the workers. One advantage is that the weld quality has improved, which means more reliability and less repair costs for them. New organizational units have been created but not at the plant level. The number of both managers and technical people needed has increased, and these positions are filled by in-house people. But no robot specific management positions have been created.

B CORPORATION follow-up interview, APRIL 26, 1981

Even for the first robot applications, management decided that this type of technology was needed to get high quality in the product. Robots were brought in to improve quality and to upgrade technology. It was not done on a cost saving basis. The first applications were part of a development process for new technology. Later applications were done for quality improvement reasons.

There is motivation throughout the company to improve the quality of products, from the very top on down. Robotics is seen as a way to accomplish this. Improved production flexibility was also an important consideration. Through a financial analysis, they saw that this programmable automation would only pay off after the second model change, but they decided to go ahead with it anyway. They are adding new models and making face lifts without buying new machines.

They knew they could achieve the quality improvement. They saw other auto makers in the United States and in Europe doing it. They knew what the equipment could do, and what problems they were having. They did not try to quantify quality in the conventional financial framework. They just went ahead with these systems. The spokesman said it is just something they have to do to get the quality level.

Now they say they are the leaders in some aspects of product quality. If all of their earlier decisions were based on cost savings, robotics use would have been greatly reduced. They would have had to look for other ways to improve the product, and none of the alternatives seemed to work very well.

Top management is committed to robotics. If they were strictly going by dollars they would have had a much more difficult time in developing these applications. Another transportation manufacturer is spending billions within the next few years on new tooling. He said this is not based on cost saving alone. Competitive reasons are very important. They want to catch up with the Japanese and with B Corporation.

The reason the Japanese are so far ahead is that they are willing to invest in a project which may take up to ten years to justify. In general, the United States is lagging in quality partly because of the use of conventional, short term techniques for justifying our expenditures.

They are also getting people out of work that is tedious and monotonous. People are absorbed into other jobs. Between reabsortion, and the attrition rate, there has not been a problem with labor.

C CORPORATION, April 21, 1981

C Corporation manufactures transportation equipment.

DIVISION CA

Division CA produces major component parts for C Corporation. The major motivations to use robots in this division is to save on labor costs. Increased production flexibility is not a major concern. The idea to use robots originated in the plant.

Most of the machines are dedicated. They do not require reprogramming. In many applications, robots represent a type of overkill, since their general-purpose capabilities exceed what is really needed. In most applications, they do not need a high-technology robot.

The need for robots evolved out of problems with hard automation. In this particular parts handling application, robots were needed because the hard automation could not handle the job.

Robots are evaluated by a comparison of time savings and cost avoidance analysis. This is more direct than return on investment.

Eight new, sophisticated robots were installed at once in a new plant that was producing a new product. No one had worked with this type of robot before. There were severe startup problems because everything was new and there was a lack of commitment on everyone's part to the bring the system on line. This included both maintenance and operations people, as well as management.

In the early training phase, there were big problems because they could not get people dedicated to the training. Management would pull people in and out of the training course. Once they completed the course, and were assigned to the robot, they were often rotated after a short period.

This problem with training seemed surprising since they apparently have the highest percentage of skilled trade workers of any plant in the division (if not the corporation). There were training problems because people were shifted around too often, and too much was started up at once, not because of people's basic ability to master the necessary skills.

It was said that when robots have been retrofitted into existing facilities, there were far fewer problems with startup and retraining. Some of the union people do not want to work with the robots because they think they eliminate jobs. On the other hand, they insist that they be the robot repair people and operators.

DIVISION CB

Division CB produces small components parts for C Corporation. Four materials handling robots were installed in 1974, and today they have nine robots (five pick and place, and four for die casting).

The idea to use robots originated within the plant. The main motivation was to reduce direct labor costs at plant level. Initially, they based the robot purchase decision on a payback analysis. Now a return on investment calculation is done.

In the six years that they have been using robots, they have not changed their financial analysis dramatically. They still quantify the same basic variables as when they started robot cost and direct labor savings. As they have gained more experience with robots, they have become more cognizant of indirect costs associated with applications engineering and training skilled trades people. They have not made any attempt to quantify these costs in their financial analysis as of yet, although they are more aware of their existence.

For them, the benefit side of the coin is fairly well defined. It is measured in terms of direct labor saving. This is due to their particular applications of robots. They are integrating them into existing automated systems. The robots are not the controlling factor on output rate, so their use has not resulted in an increase in output. Material handling robots do not affect product quality. They are still being used for the same activity for which they were originally installed, so production flexibility has not been an important concern. So the benefits are well defined in terms of cost saving resulting from reduction in direct labor.

As they continue to use robots, he thinks that they will learn more about indirect cost, and not so much about indirect benefits, at least in the present applications. He realizes they have missed many of the intangible indirect cost associated with bringing robots on-line, and maintaining them. He thinks they should be placing more emphasis on the cost of maintaining robot systems. Up to this time, the cost of maintaining the robot systems, and of training people has been ignored in their cost justification. They will soon begin to use historical information on indirect costs in future decisions.

Their maintenance requirements were higher then expected, but not because of robot reliability. Most of the problems occurred at the robot machine interface. The robot cannot compensate for many things in the way that a human can. For example, they cannot fix jammed parts in the line, so a maintenance person has to go in and do it.

They do not have NC equipment in the plant, aside from a few machines in the tool room. Robot maintenance was assigned to their electrical control systems people. They have been able to pick up the necessary knowledge to maintain robots without any problems. They have sent the appropriate hourly people to school to learn how to operate the machines. In short, training their hourly workers to use and maintain robots has not been a major stumbling block. His off-the-cuff feeling was that it is easier to teach someone to use (although not necessarily maintain) a robot then an NC machine tool.

DIVISION CC, April 10, 1981

The Division CC plant has had robots in use for the past twelve years. In 1976, there were twenty-six, but presently, there are twenty-two in use. The major motivations for their use are: man-power savings(productivity), consistency of weld, and no absenteeism.

The decision to install robotics is made at the corporate headquarters, although it does require mutual approval between headquarters and the plant. A study of payback period is done and the workers are consulted when deciding on implementation.

C Corporation does have an automation committee studying robotics. The overall impression of robotics at C Corporation is positive. They are attempting to install robots where they can replace distasteful jobs, such as painting applications.

One problem that has occurred is that there has been more down time then expected. Each occurrence usually only amounts to a fraction of a minute though. Maintenance of the robots, however, has been the major negative impact of robotization.

E CORPORATION, April 13, 1981

E Corporation is a foreign-based manufacturer of transportation equipment with a plant in the United States. They purchased two robots in the summer of 1980 and still have only two. Although they have no big plans for robots until maybe 1985, they are optimistic about robots within their existing plant. Their motivations are to reduce labor costs and to improve reliability of the production process. The decision for implementation is made on the plant level following a proposed cost and benefit analysis. The only time a corporate OK is needed is when the plant needs financial support for the project. These robots are being used on a test basis for spot welding. The tests will help them compare the robots made by two different manufacturers. They are trying to decide which to buy.

One of the biggest questions is what to do with the displaced workers. The spokesman suspects a small layoff along with an addition of skilled maintenance people. For the skilled positions, they would probably have to hire from outside the plant because he feels it would be too difficult to retrain the present workers, as a training program could take up to four years. The plant group which will maintain the robots is the welder's equipment maintenance and repair group.

F CORPORATION, April 23, 1981

F Corporation is a diversified manufacturer of capital and consumer goods. In the first applications, the novice robot users only look at saving direct labor cost in their ROI analysis. Inexperienced users only pick up on direct labor cost because they do not know what else will be affected. The experienced users like ourselves have learned that improvements in product quality and productivity, and reduction in indirect requirements CAN be quantitied, and are worth dollars.

After two years of robot use, we saw the light. We showed that benefits extended beyond labor savings, and that we could quantify it. Today, we are still looking for a three-year payback. If you can get a three-year pay back on a leading edge technology like this, you have it made.

We need management that has guts enough to say there is more potential than the direct savings the financial people will give us credit for. The management that had enough guts to justify their robot applications on the basis of longer run savings are now sitting on top of the world.

After four years of evaluating robot applications, we have not changed the basic system of financial evaluation much, but we have a much better idea of the real costs and benefits. We are convinced we can put dollars on improved quality, product life, and savings in indirect materials. As a side comment, sometimes the productivity increase is so high we can justify the robot without removing direct labor.

Four years ago, people within our plants said that robots did not meet their needs. Today, these same people are installing three or four at a time. It has taken over three years to build management support for robots. Now every management level supports robots. In my view there is no excuse for not putting robots into our plant.

All the novice users will go through the cycle of viewing their initial applications strictly in terms of direct labor cost, and then expanding their estimates of the benefits. They may go through this cycle faster then we did, since there is more information available today. You can read about robot applications in books. You can buy guides which compare industrial robots whereas, we had to write our own guide four years ago. Today, people are starting with a larger information base.

F CORPORATION: Division FA

FA Division is a potential user. Their motives include reduced labor costs and increased efficiency. He also added that there is a safety factor to consider when dealing with their products which is a motive for F Corporation because of OSHA regulations. They are presently considering robots for packaging their products. Estimates are that three or four people will be displaced, yet they will not be laid off. Their plans are to stop replacing retirees until they eventually work down to a minimal labor force. The decisionmaking process is carried out by the engineering staff of the home office which is involved in all production design projects and decisions.

G CORPORATION, March 11, 1981

G corporation is a large manufacturer of capital goods. The main motivations for using robotics were to relieve workers of unpleasant and dangerous jobs, and to "keep up" with the Japanese. He also stated that these motivations only apply for the first generation of robots and that the next generation will be used to improve product quality and to reduce labor costs. He felt that the productivity problem, the technological advances with microprocessors, and the shortage of skilled trade workers are the main motivations on an industry-wide scale.

He said that every new plant will have a robot in it where applicable. They will also install robots into existing plants. He suspects that the reduction in setup times will be a positive impact in the future.

The labor force was notified at least six months in advance of any installations. They have not experienced any safety or worker psychological problems due to robotization. No workers have been laid off, nor will any be laid off in the future because of robotization.

When positions were eliminated due to the introduction of robotics, the workers were retrained for other positions within the plant. The new positions created by robots were filled jointly by displaced workers and other in-house people. There has been a slight increase in the skill level required by the new positions. The worker's job satisfaction level has increased because many difficult, dangerous, or unpleasant jobs are now much easier, safer, and less tiresome with the use of robots. He stated that workers should not fear loss of their jobs because, within G Corporation, there is a labor shortage, not a labor surplus.

There have been new organizational units created by robotics. Two such units are a corporate planning group and a robotics group. He said that the number of managers required due to robotization has greatly increased, although no new robot-specific management positions have been created.

H CORPORATION, March 10, 1981

H corporation is a diversified manufacturer of capital goods with six robots in use during 1976 and twelve in use at the present time. The spokesman projects that they may install as many as fifty additional units in the next two years.

The most important motivation for installing robots was to reduce labor costs. Other factors that H Corporation deemed important were: to increase output rate; to improve product quality; and to reduce material wastage. Moderately important were: to comply with OSHA and to increase flexibility. He saw the reduction of labor turnover problems, and the reduction of capital costs as being unimportant factors. He thought increasing productivity should have been included on the written survey as a motivation. Presently, their corporate headquarters has a positive impression of robots and their capabilities.

H Corporation's policy requires that all capital expenditures be evaluated according to a set of standard criteria. Robot implementation decisions are evaluated in this standard method. Some of the techniques used are: calculation of payback period, return on investment, lease versus buy, and a host of other standard financial management techniques.

The major effect of robotization on labor was positive. H Corporation experienced no labor problems,psychological or otherwise. No employees were laid off; rather, they were displaced to new jobs. He indicated that no employees will be laid off in the future due to robotization. The workers accepted robotization because many of the times unpleasant jobs were being abolished.

The effects on production were also positive. He cited reliability as an extremely desirable quality added to the production process by robots. Some robots were installed in one of their plants in 1968 and they ran for two shifts a day, seven days a week, for seven straight years with no problems. He also saw robots being very useful in handling large bulky parts. He added that actual productivity was even greater than what they expected.

When asked if there were ever any negative effects of robotization at H Corporation, he responded that back in the late sixties they had a few safety problems but they were corrected immediately. He did not see any other bad side effects of robotization.

He was asked if there are so many easily identifiable benefits to the organization by installing robots and so few drawbacks, why do they have so few robots? He stated that the reason lies with management views of robots. Many managers are resistant to change of any kind if current operations are satisfactory. Conservatism rules in the business world and especially at H Corporation. When deciding how to use capital, risk is valued as undesirable. That was the major stumbling block for robotization in their organization and in American business in general.

DIVISION HA, April 20, 1981

Division HA has four plants using robots. However, only one of the four is thoroughly knowledgeable about the internal workings of robots, to the point where they would repair integrated circuit boards without shipping the components back to the factory.

The plants that do not use robots think they are sophisticated black boxes that are too complicated to maintain and are unreliable. The plants that do use robots do not have this view. To counter this reluctant attitude among nonusers our division is sponsoring robotics and automation workshops. They want to develop the analogy between NC machines and robots.

All plants are being asked to draw up a five-year plan for the use of robots and automation. If their is not a strong motivation within the plant to upgrade technology, then the division will try and promote this idea. In essence, there is a concerted effort within the division to promote robot technology, and to get the plants interested in using it.

From a maintenance point of view, a robot can be viewed as an NC machine tool. Their NC maintenance people have to be able to make the transition easily. Their NC people know how to program, and how to troubleshoot. After the two week training program given by the robot manufacturers, they are knowledgeable about robots. However, if someone with little machine repair experience were picked off the street, and asked to train to repair a robot, the transition might be difficult. In plants where they do not have experienced NC personnel, it is harder to draw on people to train to maintain and operate robots.

At present, he says the manufacturers can accommodate his needs for training his people to use and to maintain robots.

I CORPORATION, April 22, 1981

I Corporation is a manufacturer of appliances that first began considering robots in 1968-69. The first application was conceived and implemented at the plant level. The decision was made on a strict ROI basis, looking at labor saving only.

The robot did the job satisfactorily, but then the job changed, and it was not necessary to run the machines full time. The robot was no longer deemed necessary, and was taken off-line, and left idle. The machine was not utilized for a long while because no other engineers within the plant or division wanted to invest the time to get it operational for a new application.

Up to now, a robot was bought for one specific job. If the job changed, the robot was taken off-line, as illustrated above. It would not be used again until a new sponsor was found within the plant or somewhere else in the division.

If a production engineer wants to use a robot, he has to make a request for it. He says they are short of the type of people who want to develop the application and install it. He says people are too busy putting bandages on old equipment. There is little time to think about using new technology. The good fellows who would be the types who could use robots effectively are too busy making sure that the old equipment is working. They are not able to plan for the future. He said, "With American Industry running lean and mean, entrepreneurs are few and far between."

Until very recently, all purchases were done on a strict ROI basis. He said, "We were very bottom line oriented." Last year, the vice president of the division said we need more robots in order to test out if they really had a place in our operations. After that announcement, the number of applications for robot use increased significantly. The request did not specify the large ROI's that they looked for in the past. Those who really wanted to work with flexible automation went back and redid their analysis to make it look more appealing. They went ahead with several of the requested applications anyway, even though they did not have as large an ROI as they would have required in the past.

There were no major problems in getting the maintenance and operations people to get used to robots. In his view, robots are mostly hydraulic systems and electric motors. The type of machines they are using have simple logic devices. The maintenance people can learn to do this without a great deal of difficulty.

J CORPORATION, April 23, 1981

J Corporation manufactures service industry equipment. They are currently using three robots, two of which have been in the die casting operation since 1977. The motivations for using robots comes from the operational level within the plant. This has been the case for both existing and newly planned applications. Corporate headquarters is lean in technology. They have not pushed the motivations because they do not have the technical background in process improvement. They have not resisted attempts to use robots, either.

For the die casting application, the motivation was to replace an unpleasant job where there was a labor turnover. A simple, somewhat fragmented payback analysis was done. New robot applications are being planned. There is a strong push by financial people for better substantiation of cost and savings. Now they are starting to use a discounted cash flow analysis, and are looking for an internal rate of return greater than 25 percent.

They still only quantify direct labor savings. They will not modify the categories of costs of benefits based on their initial experience. Product quality and increases in productivity will not be quantified in the next round of financial analysis, but they should be quantified sometime. "As we look into the the future, these issues will have to be addressed."

K CORPORATION, April 23, 1981

K Corporation is a mass producer of small high-precision consumer products. At present, one robot is being used for welding applications. They are conservative in their approach. They do not view robots as a curiosity. They will not use them unless they are economically viable. The one robot now in use does not cause much reaction from the workers because they are already highly mechanized.

They have designed their own dedicated production and assembly equipment that suits their needs. They are producing high-volume items which have a large number of very small parts. Today's robots are not applicable to this type of work. They cannot handle the small parts, or the large volumes. They are in a totally different arena then the automotive or appliance people. One application where they do use a robot is to do welding on a product possessing few parts and low volume.

They do not have plans for robots other than the one application where it is now being used. They view their welding application as a trial balloon. They only run parts through the robot welder when it is economical to do so. As a result, they are gaining robot experience at a relatively low rate.

L CORPORATION, April 21, 1981

L Corporation manufactures component parts for airplanes. The primary motivation to use robots has come from the corporate level. They have said to use robots, and it is for the people at the plant to figure out how and where. Until recently, management gave no encouragement for a robotic program. The spokesman commented that in general, management wanted to impress outsiders that they were innovative, but internally did things to discourage innovations. For example, people would be penalized if a new development program did not work out.

Now the word has come out to "Do something about using robots." He even has robot quotas to fill. However he must still be cautious because of a perception that "The first application better be darn successful."

There is a problem with getting appropriations approved. He has to do an ROI analysis at the plant level. The corporate financial officers only look favorably upon an ROI that shows a fairly fast payback. He says this unfortunately ties in with direct labor rates. ROI will not be favorable unless there is a dramatic decrease in direct labor. He says they are trying to use robots in hostile environments where there is a problem with dust or fumes. There are problems with working conditions and with labor turnovers in these types of applications. Unfortunately, they do not show the level of direct labor savings the financial people are looking for.

The corporate financial people do not buy the argument of improved product quality either, since they say they are already making a high quality product. Arguments such as quality improvement or increased flexibility seem nebulous, and do not carry as much weight as direct labor savings. He says this type of emphasis is applied to the evaluation of all new technologies, not just robotics.

P CORPORATION, April 10, 1981

P Corporation manufactures electrical equipment. They do not currently use robots. Their major motivation for installing robotics is to reduce labor costs. The spokesman said that they plan to place robots in new plants so that they don't have to worry about the unions. He seemed overly concerned with the high cost of labor. In the decisionmaking process, an economic justification and an analysis of payback period is made at company headquarters.

Q CORPORATION, April 21,1981

Q Corporation is a primary and fabricated metals producer. Several years back the corporation started to promote within the company the idea of using industrial robots to the various plants. Today, the spokesman classifies their corporation into three categories:

1. Plants which have robot experience, and which are now mostly self sufficient in supporting their own applications. These plants have picked up the ball and run with it.
2. Plants which are not using robots, but which are pushing the technical center for assistance in developing.
3. Plants which do not use robots, and which are not really interested in doing so now.

Robot applications are based on a ROI analysis. He clearly states that there primary motive is to eliminate dangerous and unpleasant jobs. He says that there is no problem with incorporating this into an ROI framework. (Interviewer's comment: Although he did not say it, it appears the analysis is based strictly on direct labor savings.)

The robots which have been installed are less sophisticated then some of their other equipment. There is no technical problem with operating or maintaining them.

R CORPORATION, April 1, 1981

R Corporation is a large capital goods producer. Currently they have *no* robots. This includes both laboratory and production applications. In the spokesman's opinion, many of the plants are not receptive to robotics because they do not have the experience and infrastructure. They do not have the maintenance or computer system support. Using robots involves a large amount of communication between all facets of the plants, and it is hard to get them to cooperate and communicate with one another. One major division does not want to make new investments because the market is depressed, and does not look promising in the long run. Another division is concerned with keeping up in NC areas, without considering robotics. The product line in the third division is too specialized to use robots. They might manufacture 250 copies a year of a given unit, and then change the design every year. Robots are not nearly as flexible as people claim. We would prefer to apply them in fairly stable, large batch environments.

The spokesman says only the big firms will seriously consider using robotics. They are the only ones who can afford to have a robot transition staff and to finance applications. Small plants cannot afford these activities, and are less likely to get involved with experimenting.

He could estimate what jobs could be done by a robot, but emphasized that it is quite different from what jobs *will* be done with a robot. The economics will be the driving consideration, more so than the technical availability. It could be very misleading to show people the percent of jobs that could be robotized, since this might not be a very good indication of what will actually happen. Also, such estimates are difficult to make because applications are product-specific, and variables like batch size and design stability are critical factors.

ROBOT MANUFACTURER A

Manufacturer A sees the main motivations of their customers as reduction in the number of hazardous jobs and improvement in productivity and quality. Eighty percent of their customers are in the auto business. The applications include loading, inspection, pick and place, and palletizing, among others. Most of the time A has contact with main plant or corporate headquarters first. Seventy to eighty percent of their robots are placed in existing plants. As for a general breakdown of the size of the customer's plant, A finds that 60 percent of them are large and 40 percent are small facilities. The customers look at payback period most often but also at the fact that robots can run longer and are more time-efficient than humans since they do not take breaks.

A offers help in training workers to operate and maintain the robots, but it is over and above the cost of the robot. Most customers want this program. A also employs full-time instructors for this purpose.

The robot market is experiencing 35-40 percent growth per year and A is looking for 100 percent growth in unit sales from 500 units in 1980 to 1,000 in 1981. The government may influence this goal since they allow write-offs for capital equipment.

The general attitude of most customers is positive although there have been cases of a robot being taken off-line. Most of these cases involve robots that were purchased five or more years ago because the customer's needs and the types of robots available have changed substantially since then.

MANUFACTURER A follow-up interview, April 24, 1981

Novice users almost always restrict cost justification to direct labor cost. The spokesman claims that an additional 35 percent savings could be justified if indirect costs were accounted for.

As more and more companies work full time in robotics, they tend to learn more about quantifying indirect costs. He claims that novice users will expand their justification analysis as more and more information becomes available.

He says that he has seen numerous cases where users have expanded the categories of quantifiable cost savings after their initial applications. When asked how long it takes for a user to realize this, he said their was no general answer. It depends on experience.

He feels that the most important issue holding back the robotics market is the lack of trained personnel who know how to use robots and develop their applications. He points out that almost no university has an undergraduate program where students can get experience with robots. The one exception is a Macomb Community College (which focuses on apprenticeship training, not on management). Every school offers undergraduate students courses in data processing and microprocessors, but even the schools with robotics programs do not have formal programs which give students experience with robots.

Until you start educating people about the new manufacturing technology, they will have a narrow view of its applications and benefits. As a result, novice users of robots always justify their applications on the most narrow criteria until they learn more.

REFERENCES

Abernathy, William. *The Productivity Dilemma.* Johns Hopkins Press, Baltimore, Md., 1978.

Agnew, W.G. *Automotive Fuel Economy Improvement.* Technical Report GMR-3493, General Motors Research Laboratories, November 1980.

Allen, Dell K. Production-Oriented Classification and Coding Systems. In Arthur R. Thompson, Working Group Chairman (editor), *Machine Tool Systems Management and Utilization,* pages 8.8–1 to 8.8–16. Lawrence Livermore National Laboratory, October 1980. Volumn 2 of the Machine Tool Task Force report on the Technology of Machine Tools.

Argote, Linda, Paul S. Goodman, and David Schkade. *The Human Side of Robotics: How Workers React to a Robot.* Technical Report 38–8182, Graduate School of Industrial Administration, Carnegie-Mellon University, April 1982.

Arthur Anderson & Company and the University of Michigan. *U.S. Automotive Industry in the 1980s: A Domestic and Worldwide Perspective.* Technical Report, Arthur Anderson & Company, July 1981.

Beecher, Richard C. Robots in General Motors. In *Robotics: Economic and Social Implications,* pages 15–17. Congressional Institute for the Future and The Congressional Clearinghouse on the Future, Washington, D.C., Spring 1981. Report Number 3 from the Congressional Roundtable on Emerging Issues.

The Spreading Lordstown Syndrome. *Business Week*:69, 4 March 1972.

Carneige-Mellon University. *The Impacts of Robotics on the Workforce and Workplace.* Department of Engineering and Public Policy, Carnegie-Mellon University, Pittsburgh, Pa., 1981. A student project cosponsored by the Department of Engineering and Public Policy, the School of Urban and Public Affairs, and the College of Humananities and Social Sciences.

Ciborra, Claudio, Piero Migliarese, and Paulo Romano. Industrial Robots in Europe. *Industrial Robot* 7(3):164–67, September 1980.

DiPonio, John, Ford Motor Company Corporate Manufacturing Staff. Personal Communication.

Engleberger, Joseph F. *Robotics in Practice.* American Management Association, New York, 1980.

Engleberger, Joseph F. *Robots and Automobiles: Applications, Economics, and the Future.* Technical Report 800379, Society of Automotive Engineers, February 1980.

Fleischer, G.A. A Generalized Methodology for Assessing the Economic Consequences of Acquiring Robots for Repetitive Operations. In *1982 Annual Industrial Engineering Conference,* pages 130–139. Institute of Industrial Engineers and the American Institute of Industrial Engineers, Industrial Engineering and Management Press, Atlanta, Georgia, 1982.

Gizycki, R. Social Conditions for and Consequences of the Use of Robots in Five Factories. In *10th International Symposium on Industrial Robots.* IFS Publications, Ltd., Bedford, England, 1980.

Goldstein, Steven N. *Uncertainity in Life cycle Demand and the Preference between Flexible and Dedicated Mass Production Systems.* Ph.D. thesis, Department of Engineering and Public Policy, Carnegie-Mellon University, December 1981.

Groover, Mitchell P. *Automation, Production Systems and Computer-Aided Manufacturing.* Prentice-Hall, Inc., Englewood Cliffs, N.J., 1980.

Hill, John W., Jan Kramers, and David Nitzen. *Advanced Automation for Shipbuilding.*Technical Report SRI project 1221, SRI Instructional, Industrial Automation Division, November 1980. Prepared for the Naval Material Command.

Robotics in the UK. *Industrial Robot* 8(1):32–38, March 1981.

Jablonowski, J. Aiming for Flexibility in Manufacturing Systems: Special Report 720. *American Machinist* 124(3):167–82, March 1980.

Jones, David C. Choosing the Right Robot Can Mean a Lot More than Lower Costs. *Modern Plastics,* June 1978.

Katzman, Martain T. Paradoxes in the Diffusion of a Rapidly Advancing Technology: The Case of Solar Photovoltaics. *Technolgical Forecasting and Social Change* 19(3):227–36, May 1981.

Koenigsberger, F., and A.E. Debarr. *The Selection of Machine Tools.* Institute of Production Engineers, London, 1979.

Murphy, Edward S. A Preliminary Technology Assessment of the Introduction of Robots in the Motor Vehicle Industry. Master's thesis, George Washington University, Science Policy Program, 1976.

Nanda, Ravinder, and George L. Adler (editors). *Learning Curves Theory and Application.* American Institute of Industrial Engineers Inc., Work Measurement and Methods Engineering Division, Atlanta, Ga., 1977.

The New York Times, 15 November 1980.

Optiz, H., and H.P. Wiendahl. Group Technology and Manufacturing Systems for Small and Medium Quantity Production. *Journal of Production Research* 9(1):181–203, 1971.

Parsons, H. McIlvaine and Greg P. Kearsley. *Human Factors and Robotics: Current Status and Future Prospects.* Professional Paper 6–81, Human Resources Research Organization, October 1981. Prepared for the U.S. Army Human Engineering Laboratory, Aberdeen Proving Grounds, MD.

Queenen, Allen. *The Modular Approach to Productivity in Flexible Manufacturing Systems.* Technical Report MS 79–824, Society of Manufacturing Engineers, 1979.

Shackson, R, Carnegie-Mellon Institute of Research. Personal Communication.

Taylor, Theordore, Jr. *Automative Manufacturing Assessment System: Volume IV. Engine Manufacturing Systems.* Technical Report DOT-TS-NHTSA-7929, IV, U.S. Department of Transportation, Transportation Systems Center, 1979.

Terkel, Studs. *Working.* Avon Books, New York, 1974.

Toepperwein, L.L., and M.T. Blackman. *ICAM Robotics Application Guide.* Technical Report AFWAL-TR-80-4042, Volume II, ICAM Program, Wright Patterson Air Force Base, Ohio, 1980. Prepared by General Dynamics Corp. under subcontract to the ICAM program.

The Wall Street Journal, 3 February 1981. Agony Now May Mean a Brighter Tomorrow for U.S. Auto Makers.

Warnecke, H.J., H.J. Bullinger, and J.H. Kolle. German Manufacturing Industry Approaches to Increasing Flexibility and Productivity. In *Proceedings of the 1981 Spring Annual Conference and World Productivity Congress.* American Institute of Industrial Engineers, Atlanta, Ga., May 1981.

Whitney, Daniel E., et al. *Design and Control of Adaptable-Programmable Assembly Systems.* Technical Report R-1406, Charles Stark Draper Laboratory, Inc., 1981. Prepared for the National Science Foundation, Grant No. DAR77-23712.

4 APPLICATIONS OF ROBOTS IN HAZARDOUS OR INACCESSIBLE ENVIRONMENTS

In Chapter 2 we discussed the evolution of technology as a succession of extensions of human capabilities. Whereas most technological developments before the twentieth century can be interpreted as extensions of purely physical abilities (strength, speed, power), developments in the twentieth century have increasingly shifted into the arena of control and information processing. The future role of robots must be understood in this context.

The physical capabilities of robots are perhaps slightly superior to those of humans in some respects (e.g., weight lifting, stamina, tolerance of extreme temperatures), but they are certainly inferior—and likely to remain so—in some other respects, notably flexibility and intelligence. What makes robots attractive as workers is that they extend the range of man's ability to control and manipulate his environment. In a very limited sense, the robot is a substitute for a human in terms of carrying out manipulative or decision/control functions. The alternative relationships between humans and tasks are summarized by Figure 4–1. A general comment on the role of robots (which are autonomous) vis-à-vis *teleoperators* is appropriate here. In the factory environment, most jobs already are, or can be made, routine and therefore appropriate for programmable machines. Remote control of machines by

Figure 4–1.   Technology Choice.

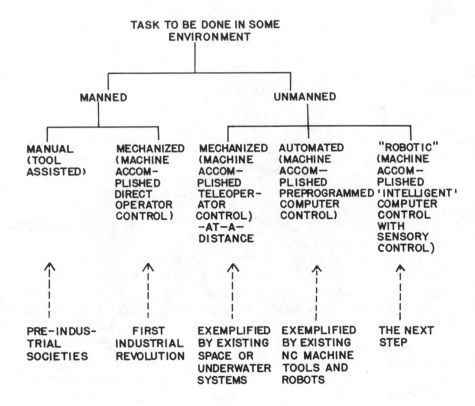

humans via teleoperators—as opposed to direct control—is appropriate only in the long run where the job is both hazardous (e.g., handling hot metal forgings or castings) and inherently nonroutine. As explained earlier, the development of advanced sensory systems will put most such industrial jobs within the capabilities of robots within a decade or so.

In the future factory environment, as discussed in Chapter 3, will be the programmed robot that repetitively manipulates tools or feeds metal cutting or forming machines, all under the control of computers. Routine decisions in production process will be made by computers on the basis of information obtained automatically by sensors. Humans will be involved in the industrial system but only in (nonroutine) functions, such as prototype design, computer programming, machine maintenance, distribution, marketing, and management.

Robots will also play a role—again, in routine functions—in nonmanufacturing sectors of the economy. Applications in underground coal mining, space, and ocean environments are discussed in some detail in this chapter. Underground mining applications are discussed first, even though robots are not now used in coal mines, because of the economic importance of the industry. This chapter concludes with briefer comments on other possible applications, such as prosthetics, maintenance of radioactive wastes, maintenance of underground pipe networks, household use, and military use.

In the nonfactory environment, routine jobs are comparatively scarcer. In fact, most jobs are nonroutine in nature. In hazardous or inaccessible environments, therefore, the "teleoperator option" is much more viable than systems where humans perform the work, and the economic competition over the next decade or two will be between remote-control systems and robots. Each will probably find some applications in the near term. At this point in time, teleoperator systems are more widely used in hostile and inaccessible environments. In the longer term, communication constraints will effectively limit the applicability of remote controlled systems in space, undersea, and underground. Virtually unlimited amounts of information can be transmitted to and from space via radio waves (e.g., pictures of Saturn), but the time lag grows very large over astronomical distances. Undersea or underground vehicles can be tethered by cable to control stations, but this obviously imposes a severe constraint on mobility. Radio waves cannot travel underwater or underground. Alternative underwater communication carriers, such as acoustic waves, limit the quantities of information that can be exchanged and also result in time lag. The only feasible way to communicate with systems deep underground is to have a series of radio wave transmitters and receivers within "line-of-sight" of each other along the length of the tunnel. Autonomous robots may be the only feasible alternative for fully exploring and utilizing the resources of space, undersea, and underground. Progress in computer and sensor technology should make robots increasingly more attractive over teleoperator systems in inaccessible environments as time goes on.

ROBOTS IN UNDERGROUND COAL MINING

Whereas most manufacturing jobs are merely dull, there is no doubt that underground mining is both unhealthy and dangerous. A miner who survives explosions, cave-ins, and other immediate hazards is still quite likely to contract "black lung" disease or some other respiratory ailment that will shorten and debilitate his life. Coal mining is historically an occupation of poor rural people with few alternatives, although union pay scales are now very high. Men who have worked in the mines do not want their sons to follow in their footsteps. The number of active miners has declined very sharply since the nineteenth century, though there was a substantial increase in the 1970s as rising prices justified expanded activity. Productivity continued to rise until about 1970 as more and bigger mining machines took their places but fell sharply in the 1970s.

Coal production in the United States peaked at close to 800 million tons in World War I. Almost all of this came from eastern or midwestern mines. Since then, it has declined more or less steadily, with a brief revival during World War II, to a low of 403 million tons in 1961. Since then, output has risen gradually. Production in 1977 was 689 million tons, of which an increasing share—61 percent in 1977—is from surface strip mines located mainly in the western states. Underground mining is declining in relative terms, though output has consistently hovered around 300 million tons per year since the early 1950s, as shown in Figure 4–2. Yet the last several

Figure 4–2.   U.S. Coal Production by Method.

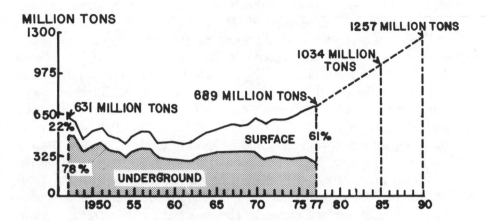

Source: President's Commission on Coal (1980).

administrations in Washington have called for greater use of domestic coal to reduce U.S. dependence on imported oil, and most long-range energy forecasts confirm that coal use in the next several decades will almost certainly increase as other fuels become even scarcer and more expensive.

Figure 4–3 shows several recent projections of coal use sponsored by the U.S. Department of Energy (Babcock 1978). Figure 4–4, from the prestigious CONAES study of the National Academy of Science (1979), makes the same point in a different way.[1] For a wide range of different "energy futures," coal is seen as the major source of energy for the United States in 2010, accounting for a higher percentage of a higher level of energy demand than is presently the case. It is unclear at present how much of this future coal production will (or can) be mined from deep Appalachian mines. The reason for doubt is clear from Figure 4–5, which indicates that surface mines in the western states currently yield about three times as much coal per workday as underground mines. The sharp decline in productivity in all mines since 1969 has actually increased the relative productivity advantage of surface mines. Trends in prices show this clearly (Figure 4–6). Wyoming and Montana coal has not increased nearly as sharply in price since 1973, as have midwestern and eastern coals. Undoubtedly, one reason for the price escalation of eastern coal is the cost of implementation and tougher mine safety and health rules mandated by Congress in 1969. The death rate in coal mining has unquestionably declined since the new legislation, as shown in Figure 4–7, although coal mining is still far riskier than most occupations.

For future reference, it is of interest to note that 45 percent of all underground mining fatalities are due to collapse of the roof supports or mine face; 31 percent to coal haulage; 14 percent to nonelectrical accidents with other machinery; and 5 percent to electrocutions. Only 3 percent of all fatalities are due to explosions and fires, while 1 percent are attributed to "other" causes. In addition to fatalities, underground coal mining is responsible for a variety of nonfatal but debilitating illnesses, especially black lung disease (pneumonoconiosis), which affects 10 to 15 percent of all underground miners.

---

[1] Committee On Nuclear and Alternative Energy Sources (CONAES)

Figure 4–3.   Coal Use Projections, Various Sources.

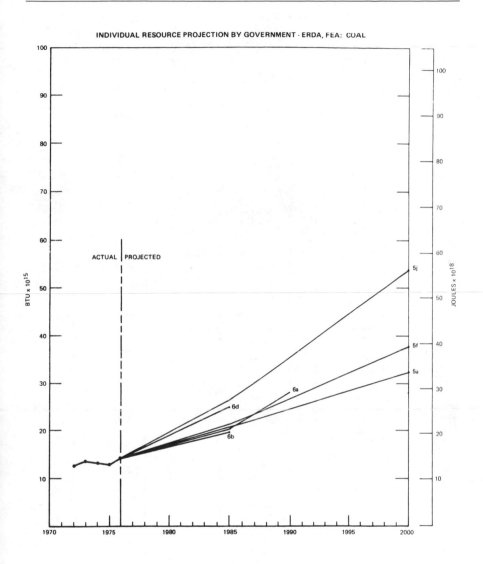

INDIVIDUAL RESOURCE PROJECTION BY GOVERNMENT - ERDA, FEA: COAL

[5a]*ERDA, 1977 Scenario SF1*
[5f]*ERDA, 1977, Scenario SF7*
[5j]*ERDA, 1977, Case 3D*
[6a]*FEA, 1976, Reference Case*
[6b]*FEA, 1976, Conservation Case*
[6d]*FEA, 1976, Electrification Case*
Source and explaination of scenarios: Babcock (1978).

Figure 4–4.   Energy Use Projections by Primary Fuels

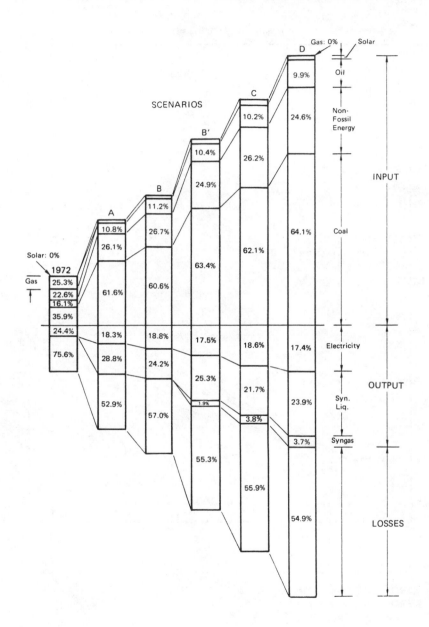

Source: National Academy of Science (1979).

Figure 4–5.   Productivity of Coal Mines.

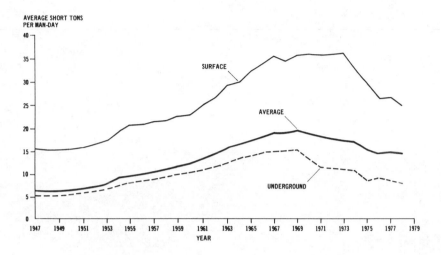

1948–1978	Annual increase rate of	(+) 2.8%
1960–1978	Annual increase rate of	(+) 0.9%
Since 1969	Productivity decreasing at annual rate of	(-) 4.1%

Source: President's Commission on Coal (1980).

Figure 4–6.   Coal Prices.

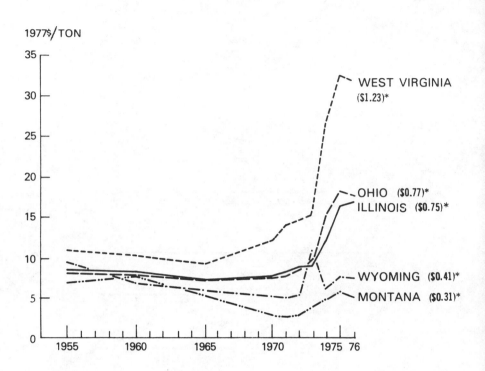

1977$/TON

WEST VIRGINIA ($1.23)*

OHIO ($0.77)*
ILLINOIS ($0.75)*

WYOMING ($0.41)*
MONTANA ($0.31)*

* 1976 Btu-adjusted price (1977$/million Btu) assuming
  State average Btu contents:
  West Virginia   13,000 Btu/pound
  Illinois: 11,200 Btu/pound
  Ohio: 11,300 Btu/pound
  Wyoming: 9,000 Btu/pound
  Montana: 8,300 Btu/pound

Source: President's Commission on Coal (1980).

Figure 4–7.   Coal Mine Fatality Rates.

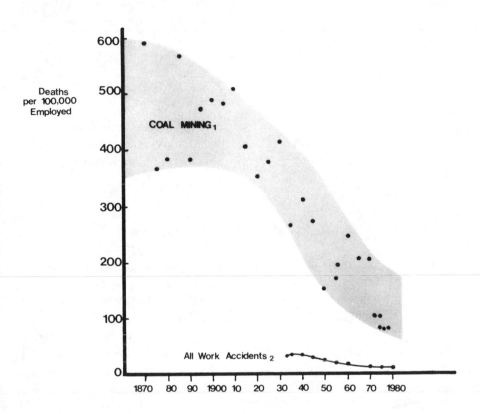

Sources: 1) Calculated from information in Historical Abstract (1961: Tables M259, M102, and M103). 2) Data taken directly from Statistical Abstract (1979).

Why not shut down the underground Appalachian mines completely and rely on cheaper western fuel? Probably, this would happen at some price differential. However, eastern coal is generally higher in BTUs per pound, lower in ash, and superior for metallurgical purposes. Also, it is near the industrial upper midwest. Lack of water (among other factors) would make it impractical to transplant a significant amount of heavy industrial activity to the Wyoming-Montana region. Thus, either the coal itself, or gas or electricity manufactured from coal, must be transported to places where it will be used industrially. Nevertheless, underground mining in Appalachia is in trouble economically. The solution widely foreseen in the industry is still more automation.

Conventional coal mining technology

Mining methods are classified according to mine layout and the technique for removal of coal from the face. The three current methods are conventional, continuous, and longwall. Both conventional and continuous mining rely on unmined pillars of coal to support the mine roof. These pillars contain 40 to 50 percent of the coal resource. The resulting mine layout is referred to as "room and pillar." In longwall mining no pillars are left to support the roof, allowing it to collapse immediately after advance of the mining equipment. High coal recovery rates are possible with longwall mining, but subsidence must be tolerated. The three basic mining methods are described in more detail in the following paragraphs. The discussion has been organized by operation, rather than mining method, since introduction of robots or other automation would not necessarily mimic the sequence of operations for conventional, continuous, or longwall mining. However, an automated mine would probably use some combination of existing operations.

The traditional technique for removing coal, and the one used in the conventional mining method, consists of three operations: cutting, drilling, and shooting. Cutting, or undercutting, is the process of creating a horizontal void at the floor level under the coal to be removed. Once a hazardous task accomplished by a prone miner with a pick, undercutting is now done by a machine resembling a large tractor-mounted chain saw. Cuts are from 7 to 12 feet deep. Coal above the undercut (or kerf) may then be separated from the seam by an adequate shock. The shock is provided by detonation of charges inserted into holes drilled above the kerf. Note that mine conditions may require variation in the cutting, drilling, and blasting operations. For instance, several kerfs in addition to that at the floor may be required to separate thick seams. A cut at the roof may even be rquired to create a clean separation at a particularly weak roof. Selective cutting may also be used to segregate unwanted material. Cuts may be made into adjacent rock strata rather than into the coal seam to reduce the production of coal dust, and vertical cuts may be used to provide uniform walls in permanent haulage ways. The objective of location selection is to remove the coal completely while generating a minimum of coal dust.

The location of holes for blasting is established by experimentation and experience in each mine. Blasting is normally accomplished with nitroglycerine-like explosives detonated by blasting caps. Explosives are packed into the hole with stemming clay or water-filled cartridges. While electronic detonation is commonly practiced, a chemical detonation device with some possible safety advantages is now available.[2] Blasting with compressed air or carbon dioxide cartridges is also used. In both cases, the cartridge is sealed by a rupture disk that bursts when the pressure is increased to a critical point by a compressor (for the compressed air) or a resistance heater (for the carbon dioxide).

Before blasting, the cutting machine has been removed from the coal face. The area is clear, therefore, for the operation of a coal loading machine which rakes the coal lumps onto a conveyer followed by hopper cars that transport coal to the mine mouth or intermediate transfer stations.

Continuous mining forsakes the cutting, drilling, and blasting operations in favor of a continuous mining machine which uses some form of a toothed rotating drum head to tear the coal from the face. The cutting head elevation and orientation, as well as the rate of advance, is controlled by the machine operator. As coal is removed, a loader, which is an integral part of the machine, rakes the coal onto an attached conveyer for eventual

---

[2] The Herendet nonelectric delay blasting system, Hercules, Inc., Wilmington, Delaware.

transport to the mine mouth. Variants on the machine are numerous and will not be discussed here.

As previously mentioned, the continuous miner and similar devices create a smaller coal lump than that produced by conventional mining. As a consequence, the amount of methane released and dust created is greater than for conventional mining. Modern continuous miners are often equipped with dust control devices. Another effect of the coal removal method employed by these machines is the high energy consumption (up to 600 horsepower) compared to conventional mining. Of course, production rates are also higher. However, even on a unit production basis, the energy required for continuous mining exceeds that required for conventional mining by an order of magnitude. That some of this energy is returned by reduced preparation costs is true, but given the practical difficulties involved in providing power in the mine environment, the tradeoff may not be worthwhile from the energy consumption perspective. Continuous mining is instead recommended by production rate and safety-related factors.

Coal extraction by longwall mining is also accomplished by toothed rotary cutting heads. However, in longwall mining, the room and pillar layout is abandoned. The milling or shearing drums are moved parallel to the working face, which may be over 600 feet wide. Coal removed as the shearing machine traverses the face falls onto a conveyer that is also parallel with the face. Coal is thus removed to haulage ways at the ends of the longwall where another conveyer system moves the coal to the main haulage system. The shearing machine may be as long as 24 feet, weigh 30 tons, and consume 800 horsepower. An alternative longwall system uses plows rather than the rotating drums of a shearing machine. A longwall plow may remove 2 to 6 inches of the face at a pass. The plow is a fixed array of cutting teeth with only a single degree of freedom (displacement parallel to the face). While plow use has been restricted to certain softer coals, some new plows are capable of extracting even very hard seams. Plows are often used in thin (42-inch) seams where shearing longwall machines are impractical. The plow results in larger lump size than the rotating drum shearer and correspondingly less dust and methane production.

The transport system includes face haulage for removal of coal in the immediate vicinity of the working faces temporary secondary haulage for removal of coal, and delivery of equipment and personnel between the face haulage system and the mainline haulage system and the mainline haulage which is a permanent and essential feature of the overall mine plan. The most common basic elements of these systems are shuttle cars, conveyers, and rail cars.

Shuttle cars are commonly used for face and secondary haulage in conventional and continuous mining operations. They are electrically powered 4- or 6-wheeled trucks that are loaded by the continuous miner or loading machine. Shuttle cars include their own conveying system which is used to distribute the payload and to unload. A tractor-trailer type of shuttle car is also available. Unloading the trailer is accomplished by conveyor, dumping, or hydraulic ram. These rubber tired shuttle vehicles offer considerable flexibility since fixed rails or frames are not required. However, for an efficient continuous operation, they are rarely the most desirable system.

The increased use of continuous and longwall miners has encouraged the use of conveyors—the continuous operation of which is more compatible with continuous extraction techniques. The face haulage system of a longwall operation employs a chain conveyor that pulls paddles along the steel conveyor channel with a continuous chain. The paddles move coal to the conveyor end where the secondary haulage system for longwall and most conveyors employed for conventional mining are the belt type, a continuous fabric reinforced neoprene belt supported by rollers driven by motors at the discharge end. While often the most economic haulage system, conveyors do have potential disadvantages. Failure at any point in a conveyor system can shut down operations, while failure of a single rail or shuttle car may not. Also, conveyors are not generally usable for transport of men and materials to the working face, so that parallel transportation may be needed.

Rail transport is being displaced by conveyors but must not be dismissed since it offers certain advantages that may be significant for an automated system. Perhaps most significantly, rail haulage may present simpler maintenance requirements than conveyors. Repair of torn conveyer belts, for instance, seems an unlikely candidate for automation. Rail systems offer more flexibility than conveyor systems in terms of geometry. Conveyors require long, straight paths while rails can be set around curves. Finally, while the discrete operation of rail and shuttle haulage has been cited as a disadvantage in a continuous mining operation, it simplifies the quality control tasks of separating coal of differing qualities, or perhaps even rock removal during mine development. Such separation is not possible in a conveyor system.

In modern continuous and longwall mining operations, the coal haulage system is often the limiting factor in production rate, that is, the capacity of the mining machines exceeds the capacity of the haulage system. The haulage system in a large mine must have sufficient flexibility to handle not only coal removal but transport of men and equipment into and out of the mine over possibly varying terrain and complex paths.

Roof support refers to measures taken to avoid collapse of immediate roof—material between the opened mine and overlying rock strata. Except in a few fortunate circumstances where solid rock lies directly above the coal seam, immediate roof consists of stratified shale, clay, and some rock that lacks the strength necessary for self-support. The term "roof falls" generally applies to the collapse of the immediate roof. Support of the entire mine overburden or main roof is practical only by the use of room and pillar mining, which leaves about 50 percent of the seam intact to support the main roof. With rising coal prices it may eventually become practical to provide a massive substitute support allowing pillar extraction without subsidence. Of concern here, however, is the collapse of immediate roof which is the primary cause of mine fatalities. Roof support or roof control is commonly accomplished by timbering, bolting, or installation of steel arches.

Timbering refers to the use of steel or wooden posts to support beams, called leaders, directly under the roof. This traditional technique has been largely displaced by bolting or the use of arches except for temporary support during the installation of one of these alternatives. Disadvantages of timbering include obstruction or narrowing of the passageways, especially if wooden supports are used, and increasing air-flow resistance that increases ventilation costs. Timbering does have several advantages. The wooden members often have sufficient give to yield under particularly severe loads, thereby relieving stress concentrations that might otherwise lead to catastrophic failure. Also, experienced miners use the audible cracking of wooden supports for advance warning of impending roof collapse.

Roof bolting (Figure 4–8) is the most common support system in U.S. mines. It consists of securing the immediate roof to the

Figure 4–8. Mechanically Anchored and Resin-Anchored Roof Bolts.

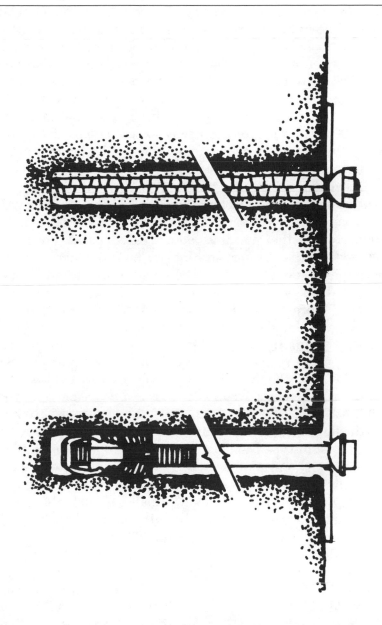

solid rock of the main roof with long steel bolts inserted into drilled holes. Typically, expansion bolts are used with cones at the upper end that expand as the bolt is tightened. Plates are used adjacent to the roof to distribute the load and secure the exposed roof. Bolting strengthens some roofs simply by binding the layers of immediate roof. Drilling and bolt installation are accomplished by bolting machines and operators. The bolting process represents the limiting step in some continuous mining operations since the mining machine must be withdrawn about every 20 feet to secure the newly exposed roof. If this were not done, the mining machine operator would be exposed to the hazards of unsecured immediate roof. Recently, the process has been streamlined by the development of miner-bolters (Harold 1980). These mining machines include bolting machines as an integral part. Thus, bolting can be accomplished more or less continuously as the machine advances. Separate operators are required for the bolting machines or bolting machine portions of the miner-bolter. Bolts are installed according to a preestablished pattern (usually one on each corner of a 4 × 4 foot square) that must be approved by the Mine Safety Administration (MSA) in advance of mine development or expansion.

As an alternative to expansion anchor type bolts, several bolting systems have been developed that use two part epoxies to secure the bolt. These secure the bolt to the stratum along its entire length, rather than at the end only. While epoxy systems are more expensive than expansion bolt systems, they may be worth the investment in troublesome roof conditions. A variation on the resin-type bolt is the injection of similar chemical strengthening agents above the seam before coal extraction. The resin fills voids, cracks, and other discontinuities, improving the integrity of the immediate roof.[3] Finally, a pumpable roof bolt of fiber reinforced resin that can be injected into bolt holes is under development (Breslin and Anderson 1976).

Roof bolting is not in such general use in European mines, though the rationale for its avoidance is not clear. Instead, steel arches are installed at regular intervals (2 to 6 feet). Both rigid and yielding arches are used, with the latter providing something of the stress relief advantage mentioned for timbering. Yielding is

---

[3] By gal-Baymidur Binder B-3, Mobay Chemical Corporation, Pittsburgh, PA.

accommodated by slip joints that are part of the arch design. It would seem that roof bolting is likely to be more amenable to automation than arch installation.

Automation in Coal Mining

The coal mining industry has generally been quick to adapt technical improvements accompanied by productive increases. Power undercutters and the Joy loading machine were among the first steps toward mine automation. Their successful application opened the door to the constantly evolving array of mining machinery commonly used in the modern mine. During this preliminary investigation it has been found that new concepts are rare in the coal mining industry. This may be due to the intense effort that has already been expended to find safer and more productive mining methods, or to the inherent conservatism of an old established industry.

The possibility of fully automated unmanned mining *is* reportedly widely accepted in the industry (Harold 1980). At this time, the state of the art in mining automation is still remote operations with human control and exposure to the mine environment. The first completely unmanned system in the United States was developed by Joy Manufacturing about 1960 for auger mining (drilling into a coal seam at grade). The Joy Pushbutton Miner (Figure 4–9) was an impressive application of available technology which, though not a commercial success, did operate as intended. High initial cost, limited application, and difficult maintenance are likely responsible for its demise (Warner 1960). As sophisticated as this machine was, it represented a relatively low (remote controlled) level of automation. Other examples of remote controlled mining machinery are becoming commonplace as manufacturers attempt to isolate the machine operators from the risks and discomfort in the immediate vicinity of the working face. Radio or cable remote controls for mining machines, for instance, allow continuous miners to advance 60 feet or more under unsupported roof before being withdrawn for the bolting operation. The operator stays in the relative safety of the bolted region.

The only example of current mining technology that enters the robotics region of the automation spectrum is automated rail haulage. Such systems are operating in about six mines worldwide

Figure 4–9.   The Joy Pushbutton Miner.

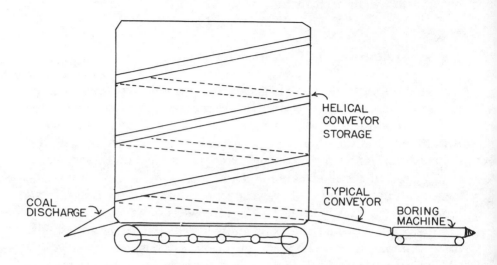

where they are used to reduce the labor requirements and risks of mainline haulage (Harold 1980). The locomotives are electrically powered and use microcircuits to perform supervisory functions analogous to those traditionally performed by operators and coal transfer station attendants. Such systems are operating successfully and research in that area is ongoing under the Department of Energy (DOE) and Bureau of Mines sponsorship. A major current effort, which is also DOE/Bureau of Mines sponsored, is the design of a system known as automated remote-controlled continuous mining, or ARCCM. Essential features of the system are described in (Arthur D Little, Inc. 1971).

"Fully automated" apparently refers to lack of in-mine operators. However, our preference is to reserve this phrase for a rather high level of self-determination—an advanced robotics system. The proposed system is a step in this direction and will require perception, selectivity, and other robot characteristics but will not be fully automated by our definition.

It is similar in concept to the Joy Push Button Miner of 1960 but operates as a continuous mining system in a room and pillar mine. The more complex geometry of this mine layout has posed challenging control problems, but the basic hardware and software for their solution have been developed. The Bureau of Mines presently foresees four stages of progressive mine automation beyond the present level based on an automated extraction system (AES). The following description is taken from a report to the bureau by Pennsylvania State University (1977).

- Stage I. The AES machine functions of cutting, loading, roof support, ventilation and dust control are individually operator performed. Face haulage is by shuttle cars. Ancillary services are the same as at present, and are substantially manually performed with limited machinery application.
- Stage II. The AES machine functions of cutting, loading, roof support, ventilation, and dust control are activated by a single operator for automatic performance. Face haulage is by bridge conveyer. Ancillary services are manually performed with more equipment application.
- Stage III. The AES machine, the face haulage bridge conveyors, and a limited amount of the ancillary services are system integrated for concurrent operation of their respective

functions. Operators would not be required to activate individual machines during straight-ahead mining but will have override capability if adjustments are required. The machine operator's role becomes one of technical serviceman to inspect the machine performance, provide necessary guidance information, and assist in machine service or repairs when the system is inoperative. Those ancillary services that are not suited to system integration will be planned for service and maintenance periods. Monitoring systems are added.

● Stage IV. A remote control mode is applied to the previous operation. The two major components of the working shift in stage IV are:

1) Production personnel absent from the immediate face
2) Service personnel present in the immediate face primarily for planned or emergency service, maintenance, turning crosscuts, and place changing.

The changes envisioned above will, of course, take place over several decades. They can be characterized generally as continued automation, systems integration, and elimination of most operator functions. Instead of a number of individually operated machines, one operator (at a remote location) will control an entire section. This technique should increase productivity while reducing the occasion for accidents. In actual fact, employment in a given mine might actually increase (e.g., from 400 to 618 employees), but tons of coal mined per man-shift would nearly double (e.g., from 14 to 27 tons). Accident frequencies would presumably decline, on the other hand, because of much lower exposure of men to areas of unsupported roof (near the face) or to machines. It is estimated that accident frequencies would drop (by stage IV) to 56 percent of the current level per man-hour and 29 percent of the current level per ton.

Robots in Coal Mines

Clearly the possible use of robots in coal mines must be assessed in the above context. In fact, as we have noted, robots will essentially replace certain categories of machine operators. They may also replace conventional rail haulage systems. In a mine, materials handling can be largely automated. One question at issue is: Can robots substitute for workers in the jobs that will remain in a stage IV mine?

The anticipated distribution of underground workers by occupation is shown in Table 4–1. It is particularly noteworthy that the operator categories, which presently account for about 38 percent of the underground work force, are expected to disappear en-

Table 4–1.   Percentage of Underground Workforce by Occupation.

Occupation	Present	Stage I	Stage II	Stage III	Stage IV
Section Crew Occupation					
Extraction machine operator	6.3	6.1	5.0	4.4	0.0
Loading machine operator	6.3	0.0	0.0	0.0	0.0
Haulage machine operator	12.6	12.2	15.1	4.4	0.0
Roof bolter operator	12.6	12.2	10.1	4.4	0.0
Ventilation personnel	6.3	6.1	5.0	4.4	4.5
Supply/utility personnel	6.3	6.1	5.0	4.4	4.5
Maintenance personnel	6.3	6.1	5.0	13.3	18.0
Subtotal	56.6	49.0	45.4	35.4	27.1
Other General Occupations					
Supply motorman	4.2	5.1	6.2	7.6	9.0
Beltman	7.7	7.5	6.2	5.4	5.5
Trackman	2.8	3.4	4.2	5.4	6.5
Wireman	1.4	1.7	2.2	2.9	3.5
Mason	1.7	2.4	2.8	3.7	4.3
Pumper-pipeman	2.5	1.4	1.7	2.2	2.8
Mechanic	5.2	6.5	7.8	9.8	11.3
Rock duster	0.7	1.1	1.4	2.0	2.3
Equipment mover	1.4	1.7	2.2	3.0	3.5
Conveyor mover	3.1	3.7	3.1	3.0	3.5
Fireboss	1.4	1.4	1.1	1.0	1.0
Oiler-greaser	3.1	3.1	2.5	2.2	2.3
Electrician	2.1	2.8	3.4	4.4	5.3
Utility man	7.4	9.5	9.8	12.0	12.3
Subtotal	43.4	51.0	54.6	64.6	72.9

Source: Pennsylvania State University (1977).

tirely in Stage IV. The question is: Will robots take their places?

The complexity of the coal mining environment necessarily encourages skepticism regarding the practicality of any scheme depending entirely on robots, no matter how intelligent. John W. Adams (Adams 1976) has summarized this complexity by suggesting that the primary characteristic of the mining environment is nonhomogeneity, and that failure to adequately acknowledge this fact has led to the poor performance of most exotic mining systems. Referring to mining in general, Adams believes that because of variations in ore quality and structural variations which affect safety, "mining probably always will require the artistry of man at the working face." He argues that productivity increases have resulted from increasingly powerful manned mining machinery and that this trend is likely to continue. He recommends, therefore, increasing mine production by increasing power to traditional methods rather than looking for a substantial breakthrough in mining methods. Adams does not address the issue of mine safety, but his arguments may be taken as representative of a mood of skepticism in the mining industry. Indeed, most mining engineers would probably argue that robots will not make significant inroads in coal mines. This conclusion might be incorrect, however, if the capital costs of the "hard" automated equipment envisioned by the ARCC program should prove excessive.

A possible alternative scenario is to utilize more conventional (and cheaper) mining equipment but to replace the human machine operators by "intelligent" robots—yet to be demonstrated. There are at present 150,000 active workers in underground mines in the United States, of which 38 percent or 57,000 are "operators." Conceivably, 50 percent of these workers could be replaced, on the basis of one robot for every three workers (two-shift operation) by about 10,000 robots. This is a sizable market that would justify a significant exploratory research and development program by the Bureau of Mines, the coal mine operators, or both. It must be recognized that no existing robot is capable of operating mining machine roof bolters or haulage cars. But it is not out of the question that future mobile robots, with sensory feedback controls, could be developed to replace humans in those jobs.

It should be remarked in passing that, though the foregoing discussion has been limited to underground coal mining, there are other underground mines that might *a priori* seem like better applications of robots. In particular, gold, platinum, and diamond mines reach much greater depths—up to 10,000 feet—and are much less suitable for heavy equipment than coal mines. While most such mines (in South Africa) currently utilize low-wage imported labor from tribal reserve areas, one can foresee the day when this may be politically infeasible. Even if the labor supply is not cut off, however, physical conditions—namely increasingly higher temperatures—may shortly rule out human labor in the deep gold mines. Robots might be their logical successors.

THE SPACE ENVIRONMENT

The space environment is obviously antithetical to living organisms. Great effort has been expended by U.S., Russian, and European scientists and engineers in creating reliable systems to transport astronauts to and from outer space and to support human life during the duration of the mission. Consequently, missions supported only from earth suffer major penalties in payload mass, acceptable extremes of operating conditions, and duration. Nevertheless, for the foreseeable future, both manned and unmanned missions will take place. Both will profit from the rapid development of automatic, teleoperated, and robotic systems that can extend the "effective presence" of people outside of protected terrestrial environments.

To date, most aspects of space operations have been controlled from the ground. This was essentially true of Apollo also, except that then human pilots participated to a degree in the monitoring and integration of the information and decisionmaking. As exploratory and other operations are conducted progressively farther from earth—for example, near Pluto, which is over one light-hour round trip from earth—the machines must necessarily act with greater autonomy because remote control of teleoperators becomes increasingly impractical due to time lags. For tasks requiring hand-eye coordination, a delay of several seconds between signal and response is simply unacceptable.[4] It is estimated that a

---

[4] According to Herbert Simon, "You couldn't run a heavyweight boxer on the moon by remote-control. He'd land on the floor at the first punch" (Bortz 1980).

remote-controlled vehicle on Mars would operate only 4 percent of the time in a "wait and move" mode, whereas with autonomous (robotic) capabilities it could perhaps achieve 80 percent operative efficiency (Heer 1981). In addition to communication time, other factors such as interference in communication links, limited number of channels, and the cost of extensive manned support will favor the use of autonomous machines. The greater the capabilities of machines in space to "sense, think, and act" autonomously, the less expensive human oversight from earth will be required. This potential tradeoff offers the most convincing case for the application of robots in the space program.

The National Aeronautics and Space Administration (NASA) has been a significant user of specialized robots for unmanned exploration of the planets for over two decades.[5] In 1962, the Mariner I flew by Venus to collect atmospheric data. In 1967, Surveyor III, equipped with an arm and a "claw," landed on the moon to perform soil experiments. In both these missions, all decisions were made from ground control on earth and radioed to the spacecraft. Once the decisions were specified, however, the information gathering and the navigation were carried out automatically at the remote site. Since 1967, there has been a continous effort by NASA, principally through the Planetary Exloration Directorate, to shift the "real-time" decisionmaking from ground control to the craft itself. The trend from remote control toward true robots is driven by the increasing cost and complexity of distant missions and the consequent need to minimize the function of ground control. NASA experience and expectations in terms of operations carried out by the craft per command and initiated by ground control are displayed in Figure 4–10. In terms of decisions made on the spot (by robots) vis-á-vis decisions made by humans, the trend is shown in Figure 4–11.

The Viking 2, which landed on Mars in 1976, and the Voyager, which probed Saturn and Jupiter in 1981-82, had roughly ten times the "autonomous" or on-board decision making capability of the first exploratory craft such as the first Mariner or the Surveyor. The Galileo, scheduled to fly by Jupiter in 1985, is designed to make roughly ten times more independent decisions

---

[5] See Heer (1978) for an overview of the history of robots in space.

Figure 4–10.   Spacecraft Automation Trends.

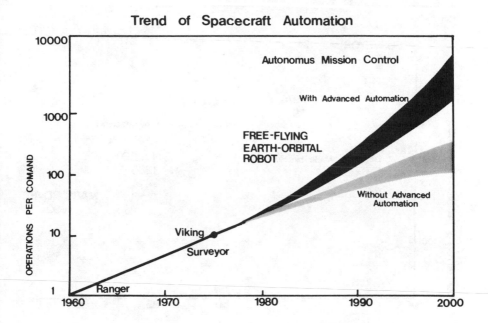

Source: Heer (1978).

Figure 4–11.   Trend of Decision Allocation between Humans and Spacecraft.

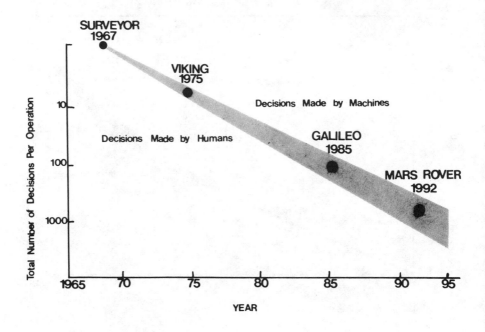

Source: Adapted from Heer (1978).

times more autonomous decisions than the Galileo. If the Mars Rover proceeds as planned, the number of on-board decisions per operation will have increased from zero to 1,000 in the 30 period from 1962 to 1992.

During 1982, NASA has deemphasized exploration in its future plans and shifted its focus more toward the utilization of space in earth orbit. The Mars rover mission, originally planned for 1985, has been delayed until at least 1990. Development of the space shuttle, space-based manufacturing activities, and plans for a space station in low-earth orbit have all been accelerated. Since most uses of space utilization to date have been communications- or surveillance-related, a more detailed discussion of possible uses of robots in the area of application follows.

Communication Satellites

The industrialization of space can be said to have begun with the Communications Satellite Act of 1962 which created the Communications Satellite Corporation (COMSAT) as the "carrier's carrier" for telecommunications within the United States. The International Telecommunications Organization (INTELSAT) was formed in 1964 to undertake similar functions on an international basis. In 1972, the Federal Communications Commission (FCC) allowed further private sector participation in the market by permitting common carriers to construct and operate satellites of their own. Western Union (Westar 1974) and AT&T (Salcom 1975; Comstar 1976) were early entrants under the "open skies" policy. The growth of the communications satellite has been phenomenal. COMSAT was profitable by 1970 and has paid increasing dividends to its stockholders since then. Its revenues in 1979 were $263 million, and it earned $40 million in profits. Numerous other satellite systems exist already or will in the near future, as Table 4–2 indicates. U.S./Canadian domestic telecommunication satellites in operation in 1978 had a total capacity of 100,000 equivalent two-way telephone circuits. This figure will have doubled by 1982. INTELSAT capacity is also growing rapidly. From its inception in 1965, it grew to include 100 countries, with 233 earth stations and 272 antennae in place by 1978 (see Figure 4–12). Total world capacity in 1978 was roughly 350,000 circuits. Capacity is expected to grow at least tenfold by the year 2000,

Table 4–2.   Communication Satellites of the World.

SATELLITE SYSTEM	TYPE								STATUS				START DATE
	REGIONAL	DOMESTIC	FIXED	MOBILE	BROADCAST	EXPERIMENTAL	MILITARY	PROPOSED	OPERATIONAL	UNDER CONSTRUCTION	IN PLANNING	R&D UNDERWAY	
American Satellite Corp*		•	•						•				1974
Arab System	•	•	•									•	1980
ATS-6 (USA)		•	•		•				•				1974
Australia		•	•				•					•	1980a
Brazil		•	•		•	•	•				•		1978
Consat/ATT (Constar)		•	•						•				1976
Consat General (Marisat)	•	•		•					•				1975
CTS (USA-Canada)	•		•			•			•	•			1975
DSCS (USA)	•		•	•			•		•				1966
ESA/Consat General (Aerosat)	•			•									1979
ESA (Marots)	•			•									1977
ESA (OTS/ECS)	•		•								•		1977
European Broadcasting Union	•				•							•	1980+
Federal Republic of Germany		•	•			•						•	1985
FltSatCom (USA)	•		•	•			•				•		1977
IMCO (Inmarsat)	•			•								•	1980+
India		•	•	•								•	1980+
Indonesia		•	•						•				1976
Intelsat IV-A	•		•						•				1975
Iran		•	•		•	•	•					•	1983
Japan-Broadcast (BSE)		•	•		•	•					•		1978
Japan-Communications		•	•			•					•		1977
LES Series (USA)	•		•	•		•	•		•				1965
NATO			•	•			•		•				1970
Phillipines*		•	•						•				1976
RCA Satcom		•	•						•	•			1974
SBS		•	•								•		1978
Sirio (Italy)		•	•			•			•				1977
Skynet (United Kingdom)			•	•			•		•				1969
Symphonia (Germany-France)	•		•			•			•				1975
Telesat-Canada		•	•						•				1973
USSR (Molniya II)	•	•	•						•				1965
United Kingdom	•		•									•	1980s
Western Union (Westar)		•	•						•				1974
REGIONAL SATELLITE SYSTEMS USING INTELSAT TRANSPONDER													
Algeria		•	•						•				1975
Andean Nations	•	•									•	•	1980s
Argentina		•	•								•	•	1980s
Denmark		•	•								•	•	1980s
Malaysia		•	•								•	•	1975
Norway/North Sea	•		•						•	•			1975

A number of other specialized experimental military communication satellite systems have not been included here due to lack of detailed information available from the respective governments.
  *Systems use other satellite transponders with their own earth stations.
  Chart courtesy of Satellite Communications
  Source: White and Holmes (1978).

Figure 4–12.   The Growth of the INTELSAT.

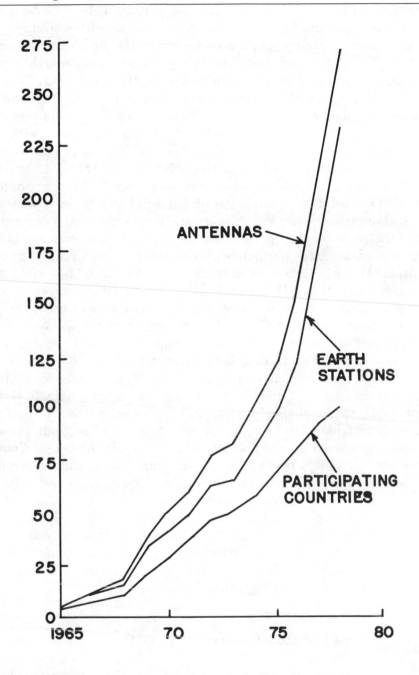

without allowing for the many new services that are likely to be introduced in the interim (White and Holmes 1978).

All existing and planned communications satellites occupy circular geosynchronous orbits directly above the equator at an altitude of 22,500 miles. Satellites in geosynchronous earth orbit (GEO) remain permanently stationary relative to the earth's surface, which means that ground-based systems can be permanently linked (by line of sight) to particular satellites. To remain precisely "on station," each satellite must be equipped with a minipropulsion system to overcome perturbing forces due to the moon's gravitational field and the oblateness of the earth itself.

The number of satellites that can be physically accommodated in GEO depends on the degree of potential interference between neighboring satellites, which is a function of the frequency band and the power of the signals being broadcast. Commercial satellites have been assigned four bands by the International Telecommunications Union. These allowed bands are $<$4-6 GH^z, 12-14 GH^z, 20-30 GH^z, and 30-40 GH ^{z}. Additions to the available bandwidth appear unlikely in view of the strong demand by other users.

For bands up to 14 GH^z, minimum separation must be 4 or 5 degrees of arc (unless television is being broadcast, in which case the separation must be doubled), which allows slots for only 70 to 90 satellites, of which less than 15 could be available to the United States or Canada. At higher frequencies (20-30 GH^z) available to commercial users, only 1-degree arc separation is greatly restricted (to date) because reception on the ground is severely hampered by the presence of rain clouds. Although clouds are only present a fraction of the time in most locations, service would be unreliable and therefore of relatively low value. Methods of improving the penetration of rain clouds by signals in the 20-40 GH^z range would provide significant system capacity increase (White and Holmes 1978). Given existing bandwidth and orbital constraints, it is extremely important to achieve high levels of utilization. Thus, availability of operational satellites on orbit at all times is of high priority.

Repair of Modular Satellites in GEO

There are presently approximately 100 satellites in geosynchronous orbit about the earth servicing as communications relays, providing weather information, performing scientific observations of the earth and objects in space, and conducting military reconnaissance. These satellites, all designed to date to be expendable, do occasionally fail. Examples are listed in Table 4–3. There are three general failure modes: wearout, random, and design.

Wearout generally affects such systems as batteries, bearings, supporting solar cells or earth directed antennae, gyros used to control altitude of the satellite, and propellants used to maintain both the satellite orientation and its preferred orbit. Generally, the degree of degradation of these systems is known and is gradual rather than sudden. There may be warning signs months before failure. Even before launch of the satellite, the rates of depletion of items such as on-board propellants are known.

Random failures can be sudden or gradual. They can affect any portion of the satellite or its supporting systems from the launch itself to gradual decay of microwave generators in the telemetry subsystem. Design failures result from failure to predict (and prevent) potential failure modes in the hardware common to a family of satellites. If a design failure reveals itself early in the life of one of the first satellites of a family, then the necessary alterations of the affected satellite components can be made on the ground prior to launch of the remaining members. However, especially in the case of communications satellites, it often happens that many of the same family are launched before the design flaw is detected. The responsible organization is then faced with watching several other satellites in a series fail in a similar manner. Less than 10 percent of all satellite failures to date are unexpected in this sense. In most cases, one has weeks to months to effect repairs and prevent the failure.

Figure 4–13 illustrates the various general approaches toward assuring reliable spacecraft operation. With the advent of space shuttle operations, it is possible to consider going into space to repair failure-prone satellites or even to retrieve satellites from some low earth orbits and return them to the ground for repair and renovation. Extensive systems analyses have been conducted to determine the relative advantages and disadvantages of the various approaches depicted in Figure 4–15 in terms of various scenarios for the number and type of satellites that will be placed in orbit during the 1980s and 1990s (De Rocher et al. 1978; Gordon et al. 1975). All the analyses suggest that in-space servicing is likely to be the most economical option. Relatively simple

Table 4–3.   Typical Failures of Communications Satellites.

Satellite	Component Failure	Type	Reparable
COURIER	Decoder	Design	Yes
TELSTAR	Decoder	Design	Yes
	Battery	Random	Yes
RELAY	Power conditioning	Random	Yes
SYNCOM	Telemetry	Random	Yes
EARLY BIRD	Fuel depletion	Wear-out	Yes
NIMBUS[6]	Solar		
Array	Design bearing	Difficult	
ATS-5	Attitude Control	Design	No
TACSAT	Structural bearings	Design	Difficult
DSCS-2	Deployable structures	Design	No
TELESAT	Power conditioning	Random	Yes
INTELSAT II	Battery	Random	Yes
	Propellant feed	Design	Probably
	Propellant relief valves	Design	Yes
	Solar array degradation	Design	Probably
INTELSAT III	Structural bearings	Design	Difficult
	Low orbit[7]	Random	Yes
	Battery	Random	Yes
	Receiver	Design	Yes
	Transponder	Random	Yes
	Earth sensor	Design	Yes
INTELSAT IV	Receiver	Design	Yes
	Thruster	Design	Yes
	Earth sensor	Random	Yes
	Telemetry beacon	Random	Yes

servicing spacecraft are possible if the primary spacecraft are designed to be repairable by replacements of failed or failure-prone modules.

Figure 4–14 depicts one generic approach that has been suggested for the servicing of unmanned satellites in both low earth orbit (LEO) and geosynchronous orbit (GEO). In this approach, an unmanned service vehicle (SV)—essentially an orbiting ware-

---

[6]NIMBUS is not a communications satellite, but had a problem that may occur on future communications satellites.

[7]INTELSAT III was injected into a low transfer orbit; hydrazine propulsion was used to achieve proper orbit.

Figure 4–13. Satellite Availability On-Orbit.

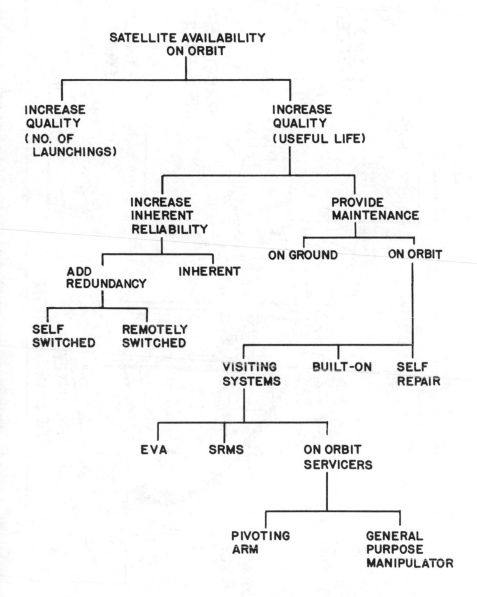

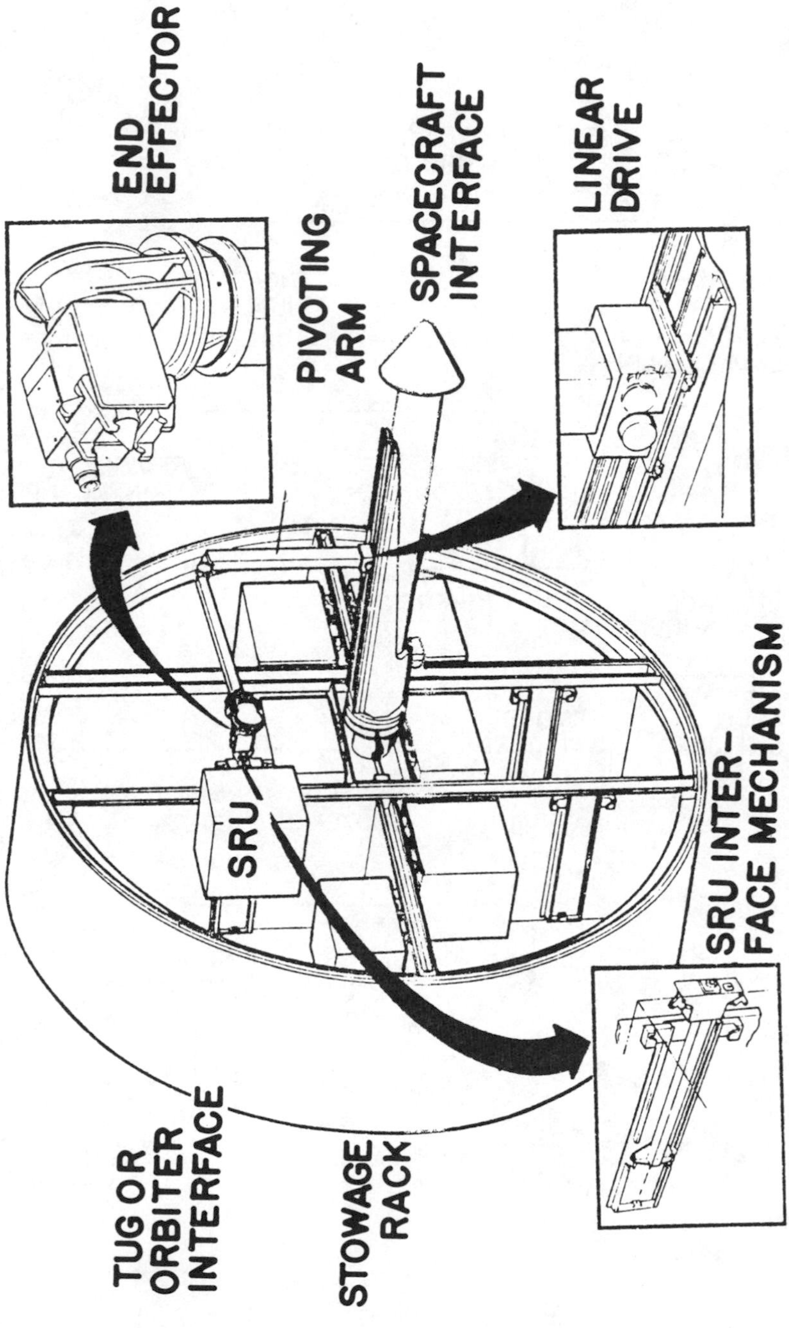

Figure 4–14.   Pivoting Arm On-Orbit Servicer.

house—is loading in LEO with modules of electronics, propellants, batteries or other devices which ground controllers have decided are necessary to restore or ensure full function of one or more satellites. These "spare parts" modules may be loaded onto the orbiting SV during a mission of the space shuttle. Actual repair missions are entrusted to one or more self-powered robot service spacecraft (SS). An SS is attached to a propulsion unit such as the orbital transfer vehicle (OTV) planned for the late 1980s and is transported to the region of space where the satellite servicing sequence can begin.[8] It rams into the docking adapter of the spacecraft at 3 to 5 kilometers per/hour and mechanically couples. The SS is equipped with an arm that can be rotated about the docking axis, translated radially to the axis, and extended along the length of the docking axis. It thus possesses 4 degrees of freedom (DOF). The arm extracts the damage module from the target satellite, stores it in a rack on the SS, and then extracts a standard replacement (SRU) module and inserts it into the proper receptacle of the satellite being serviced. The modules may be on the side of the spacecraft or, more conveniently, in a standardized location on the back of the spacecraft. The SS then decouples from the satellite and returns to the carrier or proceeds under its own power to the next craft to be serviced.

Implicit in this approach to the servicing of a wide range of spacecraft is the restriction that the hardware likely to require servicing be packaged in a modular form that is standardized within rather broad limits. Such modules must be equipped with connectors (electrical, mechanical, fluid, electromagnetic) that can be broken and reestablished by simple pulling and pushing operations and possibly augmented by releasing and reinserting clamps or similar mechanisms. Surprisingly, it is considered possible to design most spacecraft systems in this manner with a mass penalty of no more than 25 percent as compared to "mass-optimized" expendable designs. Clearly, the success of a servicing program is predicted on the design of all spacecraft to be serviced in a manner consistent with the standardized replacement procedure described above. There is resistance to this approach from the segment of the aerospace community responsible for the design of the spacecraft. It is felt by some designers that failures can be

---

[8] See (Romero 1981) for an overview of NASA's current plans for space transportation.

prevented, given the greater design flexibility permitted by weight savings. In other words, potential savings to be expected from a servicing approach must be weighted against inevitable sacrifices (in performance and inherent reliability) attributable to the limitations of a suboptimal design.

While we can expect the development of service robots with increasing capabilities to sense, think, and act, spacecraft design and repair cannot, for the forseeable future, assume general-purpose space robots that are able to solve any conceivable problem of a malfunctioning spacecraft. Rather, there must be a compromise approach in which a team of engineers very carefully considers the evolutionary path of design and operation of spacecraft and identifies how the availability of servicing robots would decrease the initial cost of spacecraft, extend their operational lives, and allow retrofitting of existing spacecraft as technology improves. It cannot be overemphasized that module replacement alone is not an adequate approach because the burden of making the system work falls on the spacecraft designers who are subject to many other constraints that may be incompatible.

The following is a list of possible research topics to develop robots for the repair and maintenance of orbiting spacecraft, and for space manufacturing in general:

1. Ascertain how much "intelligence" and decisionmaking capability is needed for a given set of tasks and, then appropriately distributing the intelligence among ground control, in-space control centers, and autonomous machines.
2. Examine assembly of spacecraft or of other systems in space. This involves a wide range of problems, including (1) attaching the servicing vehicle to the system to be repaired, (2) establishing a stable platform from which to work,[9] (3) designing large manipulators that are potentially very flexible, (4) controlling these manipulators in space, especially groups of multi-jointed manipulators working together whose actions must be coordinated, (5) designing manipulators and "tool sets" that could be used in the repair of spacecraft and other parts which are well

---

[9] To appreciate this problem, imagine a swimmer trying to tighten a bolt on the hull of a large floating ship. The swimmer would rotate, instead of the bolt, when he applies a force to the wrench.

suited to the space environment (e.g. top, locking screws rather than slot screws) and that would offer the greatest utility in servicing the needs of designers.

Once space assembly is feasible, the assembly of large satellites out of previously in-orbit or newly provided satellites and specialized subarrays could be considered. For example, an older satellite might be attached to one of the large communications platforms being proposed for the 1990s.

3. Investigate the use of "remote" modelling and simulation to solve problems. For example, a "projective teleoperator" (PT) is a two-part teleoperator in which the unit in space observes the object to be manipulated and transmits the state vector of the object and its own internal status to a controlling unit on the ground. The scene is duplicated on a ground monitor (possibly three dimensional). An operator on the ground then uses a glove or similar hand controller to direct the start of the operation in space by a continuous motion while watching the simulation screen. Computer programs on the ground and in space then project these motions ahead in time and "operations space" and continually send signals back and forth to each other to determine when the two sets of motions (i.e., projected in space and true—on the ground) begin to diverge. When excessive divergence occurs, appropriate stop and refresh signals are sent. In this manner, the time delay inherent in long distance communications is averaged out rather than incorporated directly into a control loop involving every motion.

4. Design docking systems that do not depend on large fixed receptors in one part of the object satellite but that allow attachment and guidance to other points, especially to allow for nonaxial dockings.

5. Design systems that allow for tracking and joining with a spacecraft that is not passive but may be venting gas or propellants in a failure mode or possible satellite that is spinning and nutating. This may require rotary joints on the service satellite.

The subsequent steps, as now envisioned, would be as follows. The SS decouples from the primary carrier rocket during the final approach to the spacecraft to be serviced and docks with the spacecraft using the "bayonet" spacecraft interface (Figure 4–14) technique developed during the Apollo program.

Advanced Teleoperator and Robotic Systems

The use of unmanned (robot) satellites to facilitate repair and maintenance of modularized satellites represents a reasonable near-term goal. NASA's Marshall Space Flight Center is currently studying how to use teleoperator and robotic systems to fix satellite antennas and to repair and exchange modules (NASA 1982). However, it is a limited capability. For instance, manufacturing of a large spacecraft already in orbit is not possible, access to parts deep within the structure of a spacecraft is precluded, and the adaptation of the assembly/disassembly system (A/DS) to other tasks is not possible. With the great increase in lift capability that is provided by the space shuttle, far more ambitious tasks should be possible in low-earth orbit. More ambitious operations in low-earth orbit would enhance the need for further extensions of robotics technology (e.g., to support on-orbit and lunar based fabrication and assembly), which would not otherwise be required for many decades by space programs supported wholly from the ground.

The space shuttle is expected to be capable of transporting approximately 30,000 kg of payload into low-earth orbit in a vehicle approximately the internal volume of a rail box car 18 meters long by 8 meters wide (NASA 1976). These dimensions and lift mass are consistent with the sizes and mass of machinery that could form the basis of a small machine shop and assembly station for the generalized construction and repair or remanufacturing of satellites in orbit. Eventually such a system could be deployed in low earth orbit (LEO) permanently, along with a power module such as that being planned for long-term missions of the space shuttle and most likely attached to a platform also constructed in two-earth orbit.

NASA's Goddard Space Flight Center is seriously studying the creation of a space-borne assembly/disassembly system that could be deployed during the 1980s from the shuttle or operated in the shuttle bay. A significant number of scientific satellites are already in orbit. Some of these could be brought back to operational status or improved in instrumentation if inspace modifications could be made to the systems. Very large systems such as the large space telescope to be launched in 1984 would also be amendable to repair and refurbishment by in space repair facilities. NASA intends to conduct a wide ranging study of all phases of robotic associated manipulation of parts and components appropriate to automatic manipulation, operate prototype robotic assemblies, conversion of available CAD/CAM programs to space (long data link) situations, evolution of new software for the same purposes, and the building of test units to develop experience with robotic assembly.

Figure 4–15 is a schematic of the space shuttle bay with the robot A/DS station on the right side, a robot transfer arm right of center holding a satellite that is being inserted into a capture/release gig. The 50-foot-long shuttle remote manipulator arm will then pick up the capture/release gig and deploy the renovated satellite. A second satellite left of center is awaiting transfer by the mobile robot arm to the A/DS system.

Naturally, the technical advances required for development of a space-borne A/DS system would find wide application in many terrestrial applications such as those reviewed in Chapter 3. Discussions of recent developments can be found in Korf (1981), Lund, Pinksy, and Showalter (1980), and Long and Healy (1980). Basic advances are required in several areas that are also common to terrestrial needs such as:

- Improvements in robot dexterity and sensors (especially vision)
- General and special purpose teleoperator/robot systems for materials handling, inventory control, assembly, inspection, and repair
- Improvements in computer control of large, integrated, dynamic hierarchical systems using sophisticated sensory feedback
- The embodiment of "managerial" skills in an autonomous adaptive-control system

Special consideration will have to be given to problems associated with the tradeoffs between cost of developing successively

Figure 4–15. Assembly/Disassembly Station (A/DS) in the Bay of the Space Shuttle.

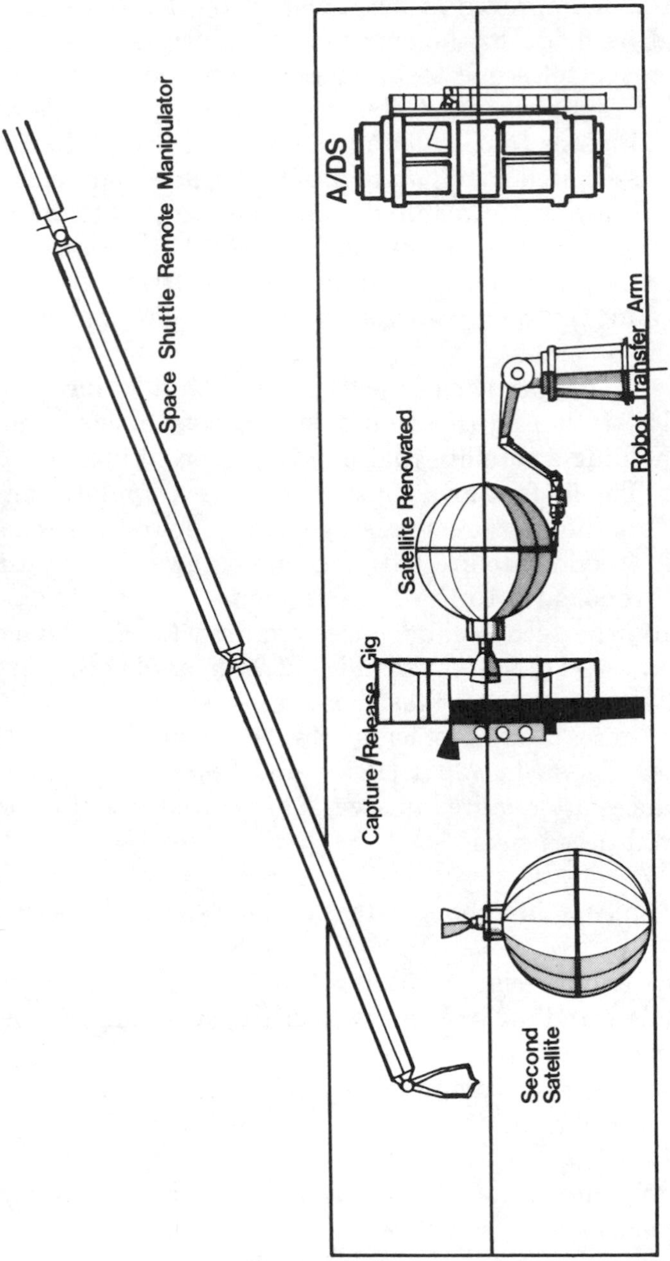

more sophisticated autonomous machines and the costs of having people in the loop. This problem is not strictly tied to the present cost of maintaining people in space because the automatic systems can also be used to reduce the cost of manned space flight in the future.

On-Orbit and Lunar-Based Fabrication

In the late 1980s or early 1990s there will be a need for the in-space fabrication of devices either from raw feedstocks such as could be obtained from the moon or asteroids or from partially prepared feedstocks shipped up from earth onboard large launch vehicles or by means of electromagnetic mass drivers. The "nonterrestrial utilization of materials" working group of the NASA/Santa Clara summer study (Long and Healy 1980) concentrated on analyzing the first steps in the development of fabrication technologies especially appropriate to the space environment. The group envisioned the initial step to be a space manufacturing facility (SMF) in low-earth orbit that would initially be highly dependent on earth for raw materials inputs, logistical support, and management support. It was assumed that one of the major tasks would be to develop ways whereby the support from earth would be progressively decreased by greater use of nonterrestrial resources and solar energy. There would also be a progression in human intervention per unit of material goods outputs from manned (with automation) support to greater use of teleoperators and eventually to robots wherever possible. It was suggested that the SMF would produce goods of use initially in space (e.g., hulls, components of robots, antennae, life support systems and fluids, telescopes, spacecraft components) and goods for terrestrial markets as soon as the value added in space manufacturing would result in competitive or totally unique products.

A survey of 220 commonly employed terrestrial manufacturing processes and of nine techniques proposed previously to the unique conditions of space was conducted to determine which techniques would be appropriate to working in space conditions (vacuum, zero gravity, use of solar energy), require minimum use of terrestrial materials (especially carbon and water in the early phases), and be amenable to automation and robotic manipula-

Table 4–4. Manufacturing Processes Applicable to Space Based on Terrestrial Experience.

PREFERABLE	USABLE WITH RECYCLING OR ADAPTATION
I. CASTING	
Permanent	Sand
Centrifugal	Shell
Die	Investment
Full-mold	
Low-pressure	
Continuous	
II. MOLDING	
Powder metals and ceramics	
III. DEFORMATION	
Thread rolling	Forging (with electrical drives)
Magnetic pulse forming	Lead-in mill
Electroforming (basalt electrolyte)	Extrusion (basalts)
Rolling—reversing mill	spinning (glass & basalt)
IV. MACHINING[a]	
Laser	Turning (basalts)
Electron beam	Drilling (basalts)
	Grinding (recycle binder, using Al_2O_3-grit)
V. JOINING	
Cold/friction welding (metals)	Metal fasteners (need fusion
Laser beam welding	preventers)
Electron-beam welding	Glues (need carbon)
Induction/HF resistance welding	
Fluxless/vacuum brazing	
Focused solar energy	
Metal fasteners (permanent)	
Stitching (metal or inorganic threads)	
Staples	
Shrink and press fitting	
VI. CONTAINERLESS	
Surface tension	Metal and/or ceramic clays
Fields	(binder loss)
E&M	
Centrifugal	
Gravity Gradients	
Direct solar heating (differential)	
Vapor deposition	
VII. CONTAINMENTS	
Powder/slab—cold welding	Metal and/or ceramic clays
Foaming (metals/ceramics)	(binder loss)

[a] In a vacuum environment most machine techniques will require a pressurized container to prevent cold welding effects.

Table 4–5. Scenario for the Growth of Space Manufacturing and the Use of Robotics

Raw Materials and Materials Processing	Manufacturing Technology	Energy	Commercial Applications
Initial raw materials and materials processing base	Ground demonstration of starting kit	250 kW solar power demonstration	Ground demonstration of teletourism
Asteroid exploration	Substitutability research	Solar power module for initial manufacturing station	Teleoperation of nuclear facilities
Automated/ teleoperated lunar exploration	Teleoperation research and demonstration	25 kW power extension package	Ecosystem control
Transition to non-terrestrial raw materials	Ground based space farm	Lunar power station for lunar raw materials base	Prosthetics
Asteroid raw materials exploration	Deployment of initial starting kit in orbit	Ground demonstration of large space mirror	Laser communication links
Complete transition to non-terrestrial raw materials	Ground demonstration of expanded manufacturing capability	Extension of manufacturing power station	Orbital teletourism
	Deployment of external tank processor	Space manufacture of solar sails	Lunar teletourism
	Ground demonstration of large space structure manufacture		Orbital and lunar tourism
	Satellite manufacturing, testing and repair	Solar power station for space needs	Construction of large space structures
	Completion of space manufacturing	Nuclear power station	
		Impulse launcher	
		Laser power transmission	
		Solar power satellite	

Source: NASA.

tion of the production processes. Table 4–4 summarizes the techniques that appear especially applicable.

Special attention was given to the possibilities of "starting kits." These would be devices that could use local materials and solar energy and build copies of themselves and a wide range of other devices of manufacturing. Figure 4–16 illustrates the production section of one such device called an impact-molder. Both this device and an alternative system based on the manipulation of clay-like starting materials would use powder materials and ceramics. One of the most important characteristics of both starting kits is the automatability of the tools included for the initial shaping and trimming operations. It appears that the forming and shaping functions are the farthest from present state of the art. However, it was felt that the starting kits would be deployed in the near term as fault-tolerant, easily reprogrammable devices of wide use in a SMF.

Table 4–5 gives one possible overview of the course of development that space exploitation might follow. Development of the key raw materials and manufacturing technologies will allow the creation of successively large energy sources and materials supply sources. These developments will, in turn, widen the possible commercial interest. Orbital and lunar "tourism" are possibilities that might not require completion of all these steps prior to implementation. Construction of large structures in space, which will certainly be required in any large-scale space program, will be expedited given general purpose construction units for space (CUS) operations.

Both the awkwardness of working in spacesuits and the desire to minimize exposure to cosmic rays (galactic and solar) by space workers promote an interest in having a general purpose assembly robot that could assist in construction and repair of space structures. A conceptual version of such a robot is shown in Figure 4–17.

We assume the CUS is capable of independent motion in the local space about a major construct in earth orbit such as a space power station (SPS). The version illustrated is for "walking" along beams equipped with two "legs," each with 4 degrees of freedom. The CUS also has 3 "arms." Two would be heavy-duty lifting or holding arms with 5 degrees of freedom and one would be a lighter manipulator arm with 6 degrees of freedom. Arms

Figure 4–16.   Automated Impact Molder Using Robots for Trimming and Inspection of Parts.

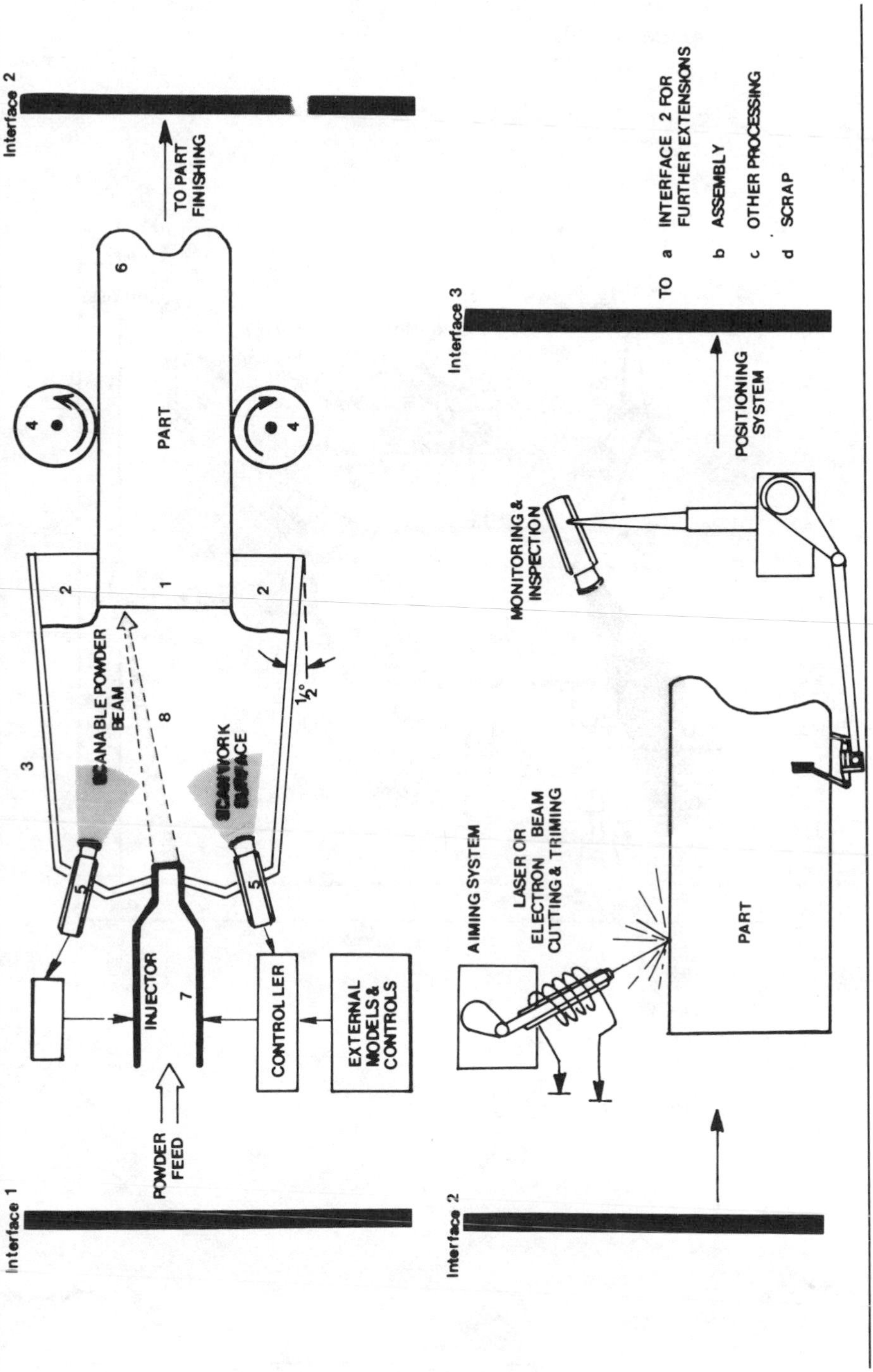

Figure 4–17.   Artist's Conception of a Construction Unit for Space (CUS).

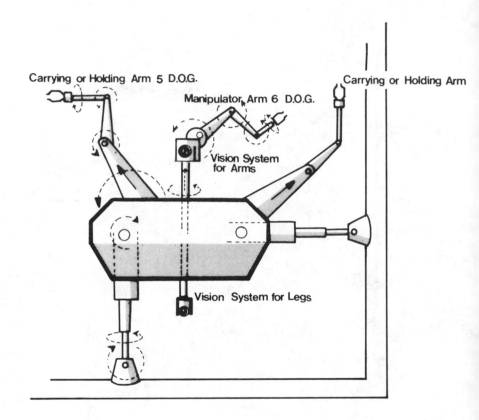

and legs would have separate vision systems. The illustrated version would be capable of working while standing on either one or two legs.

ROBOTS IN THE OCEAN

The ocean floor contains valuable minerals and fuels: oil, gas, sulfur, and potash that can be extracted through boreholes. Oil and gas are already being pumped from offshore wells in the Santa Barbara Channel in California, the Gulf of Mexico, the North Sea, the coast of West Africa, the South China Sea, and the Indonesian archipelago. Drilling is underway off the coasts of Brazil, Argentina, India, China, and elsewhere. As in coal mining, drilling operations, do not appear to be suitable for robots, but exploring the sea floor (e.g., for signs of oil seepage) and assisting with underwater operations, such as plugging blowouts, putting out fires, and inspecting and repairing offshore platforms may be important applications.

There are also subsurface deposits of iron ore, coal, barite, molybdenum, copper, and zinc. Then there are unconsolidated deposits of heavy metals and nodules of manganese and phosphorite (phosphate rock) that have been formed from seawater by a slow process of biogeochemical precipitation. Phosphates could be taken from deposits off Georgia and California with existing technology. Heavy-mineral-bearing sands exist in many offshore locations, especially submerged streambeds. One of the best known is the thorium/titanium containing "black" monazite sand of Kerala, on the southwestern tip of India. Tin-bearing gravels are found off the coast of Malaysia.

The manganese nodules known to exist over wide areas in the Pacific Ocean constitute a major resource. They average 29 percent manganese, 5.3 percent iron, 1.28 percent nickel, 1.08 percent copper, and 0.25 percent cobalt. The last three metals may be more important than the first. Estimates of the amounts of those metals suggest that in the Pacific Ocean alone they exceed the reserves on dry land by at least a factor of 5 (Arrhenius 1974) and possibly much more. Even the minor constituents—copper and nickel—are present in quantities that dwarf known reserves on land.

While major mineral deposits will probably be dredged by specialized equipment operated from surface ships, there are peripheral functions (surveillance, hazard avoidance, underwater repair, and rescue) that could well be carried out effectively by autonomous robots. Again, the alternative approach would probably be to use teleoperators.

Actual economic exploitation of ocean floor minerals has been limited, so far, to the continental shelves within 70 miles offshore and in water no deeper than 350 feet. Most activity in the next twenty-five years will remain more or less within these limits, subject to modest advances. But the technology for finding, recovering, and processing manganese nodule deposits is now in a fairly advanced stage, close to commercial feasibility. Depths of 4,000 feet have been reached, and further tests to 15,000 feet are underway. The biggest problem is cost; investment of $50 million is required to reach the prototype stage. And some $700 million is further necessary in the commercial stage, both for offshore investments in mining and transportation and onshore investments for terminals and processing plants.

Undersea Robots

It appears increasingly feasible to develop two classes of underwater devices: observing robots (ORs) for reconnoitering the deep oceans (20,000 feet) and teleoperated devices for manipulating material objects underwater at great depths. The distinction is necessary because of the fundamentally divergent physical and energy needs posed by data collection versus manipulation.

Manipulation of objects, especially large objects, on the deep sea floor will require considerable power (e.g., 5 kW) over relatively short periods of time (seconds to hours, or even days, in shifts). There may be compelling reasons to have real-time high-resolution data (e.g., television) from some sea floor sites. Real-time observations, and possibly remote control, will require a physical cable to transponders at the ocean surface. Even very thin cables (e.g., optical fibers) will exert large and variable forces on a teleoperated manipulator (TM). Considerable propulsive force must therefore be exerted by such a teleoperator device to position itself and to do work on objects. However, TMs need not have the capability for long-range undersea journeys. They could be positioned by high-speed air or surface craft. ORs have very different operation objectives and technological needs. They should be as inexpensive as possible to permit their deployment in fairly large numbers. Small physical dimensions (mass, cross-sectional area and length) are required to minimize the total energy requirements for long-range (1,500 km) cruises and minimize the expense of providing the energy sources on-board.[10] ORs must navigate to and within a preselected target zone with sufficient accuracy to allow a complete survey of the target zone in a time that is short compared to the cruise time for surface vessels, and then return to the selected home port. The OR must also be capable of leaving (and possibly recovering) acoustic beacons at interesting sites. It must also provide batch data transmission and reception with a satellite while remaining beneath the turbulent zone of the sea. ORs should also be capable of operating in coordinated groups to permit simultaneous observations of events from different locations and to relay communications.

Detailed analyses (Johnson, Venderese, and Hanson 1976) and limited experiments (Dixon, Johnson, and Slaghle 1980) indicate that a combined-function robot is now possible which would have a range of 1,600 km, limited manipulative capability, and reconnaissance capacity. Work is under way in Japan and the USSR on control of manipulators by an underwater robot (Yastrebov and Stefanov 1978). Assuming the need for 500 W of continuous electronics "overhead" power (navigation, guidance, data collection, etc.) and 250 to 350 W of propulsive power, this robot would have a diameter of 1 m, a length of 4.5 m, and a mass on the order of 6,000 kg. The payload volume and mass would be approximately 1.4 cu m and 1,400 kg, respectively. The vehicle could be powered by a lithium inorganic battery (nonrechargeable) or an 0.8 kW radio thermal generator (RTG).[11]

---

[10] Current observing vehicles are limited to voyages of just several hours in duration, with a range of approximately 30 km. Ranges of 1,500 km are required for underice exploration.

[11] There is already international concern about the use of RTGs and other types of radioactive power sources in satellites and undersea exploratory vehicles. A U.S. satellite powered with an RTG device crashed over Canada in 1979. Apparently, the Indian government expressed serious concern when the Central Intelligence Agency lost an RTG-powered device in a landslide at the head of the "sacred" Ghanges River.

It is interesting to note that with a change in "power profile" philosophy—that is, run all electronics overhead from rechargeable batteries (recharged from a very small RTG) only when the full electronics capabilities are needed—vehicle propulsion energy requirements might drop to approximately 50 W. The overall vehicle size could then decrease to approximately 30 cm diameter by 1.2 m long. Payload volume and overall mass would also decrease. An overall mass less than 1,000 kg might be possible.

Obviously, the objectives of low, cost and long-range operations indicate the smallest possible craft. This is only possible if the available power is proportional to volume. Unfortunately, with the electric battery storage, it is available energy (not power) that is proportional to volume. To overcome this difficulty is a challenge for basic research.

Development of the unmanned and autonomously controlled vehicles apparently started with a University of Washington vehicle UARS financed by the Department of Defense Advanced Research Project Agency (ARPA) in the early 1970s. This battery-powered vehicle was 3 m long and 48 cm in diameter. It was used to obtain underice profiles in the arctic over a rosette-patterned track approximately 30 km in length. Recovery was by means of a net (Geyer 1977).

R. Frank Busby Associates, Inc. (1979) has reviewed the development of remotely operated vehicles through 1978 and attempted to determine areas of research most pertinent to the application of these devices. Table 4–6 summarizes military and scientific applications. Lack of recognition of possible industrial applications is surprising and consistent with the more recent technical literature in the field of ocean engineering and exploitation.

Major qualitative factors noted by Busby are: (1) Severe problems of cable tangling and rupture exist with towed devices; (2) only 5 percent of the identified undersea tasks required manipulation; (3) at the present time, the field is so new that there are no design standards for submersible systems that could be expected to be of use to a wide range of users. This is a market development problem. They anticipate a growing need for color television transmission from submersibles, emergency locaters for devices primary navigation systems fail, development of techniques whereby the remotely operated vehicles (ROVs) can assist

Table 4–6.   Remote-Operated Vehicle Work Categories.

Industrial	Military	Scientific/Research
Swimming vehicles		
Inspection	Inspection	Inspection
Monitoring	Search/identification	Survey
Survey	Installation/retrieval	Installation/retrieval
Diver assistance		
Search/identification		
Installation/retrieval		
Cleaning		
Bottom crawling vehicles		
Bulldozing	Drilling	Inspection
Trenching	Trenching	Survey
Inspection	Installation/retrieval	
Manipulation		
Towed vehicles		
Survey	Search/identification/ location	Geological/geophysical investigations
	Survey	Broad area reconnaissance
	Fine-grained mapping	Water analysis
	Water sampling	Biological/geological sampling
	Radiation measurements	Bio-assay
		Manganese nodule survey/study
Untethered vehicles		
None at present	Conductivity/temperature pressure profiling	Bathymetry
Many inspection and exploration task proposed	Wake turbulence measurements	Photography
	Underice acoustic profiling	
	Testing prototype submarine designs	

Source: Busby (1979)

divers, acoustic imaging devices, a means for compensation of heave, and increased reliability and performance. Development of higher power thrusters and energy sources would naturally expand the applications of (ROVs).

Several operational requirements for conducting scientific tasks by ROVs were identified by Busby Associates:

- Control/maneuverability (over flat and steeply sloping bottoms)
- Pursuit/capture capability (fish, for example)
- Higher data quality, increased telemetry rates, better viewing (resolution range), color television
- Three-dimensional viewing
- Extended range of viewing (past natural viewing limits)
- Organism detection and location techniques (sonic)
- Better navigation positioning
- Minimization of sediment disturbances
- Minimization of disturbance on organism behavior
- Greater manipulative/sampling effectiveness
- Increased the duration of observations.

A navigation problem of particular interest in commercial operations and to the machine intelligence community is the navigation of a free swimming device inside a structure (e.g., the struts of a Texas Tower, a water/sewer system, or a flooded nuclear containment vessel). In the presence of metal structures, current acoustic navigation and positioning systems become unusable due to multiple echoes. Perhaps long and short baseline acoustic techniques would be useful. Alternatively, perhaps a "model" of the structure can be memorized by the ROV and "dead reckoning," plus the use of doppler sonar off known structural members, can be used for precision location inside such structures.

Recent growth of interest in this field has been spectacular. In 1974 there were only 20 known ROVs. Boller and Busby (1980) have reviewed the present inventory of ROVs (approximately 410) and noted 7 to 10 untethered vehicles among them. Bender (1980) points out that, in addition to the 7 to 10 untethered ROVs listed by Boller and Busby, the French navy operated over 120 PAP-104 mine neutralization vehicles built by Societe ECA, Meudon, France.

Booda (1979) notes that manned submersibles are still preferred in most complex tasks. However, the rapid improvement in the abilities of unmanned vehicles and their generally lower costs are driving the extensive use of unmanned units. The untethered ROVs are seen to be of greated use at extreme depths and for tasks of search and identification.

Experimental work is being actively pursued at seven institutions. Centre National pour l'Exlpoitation des Oceans (CNEXO), Paris, has tested a 3-ton survey system to depths of 6,000 m (Offshore 1979). The "Epaulard" will be used in manganese nodule surveys. Sea trials are scheduled by workers at Heriot Watt University for the ROVER in 1983. ROVER will operate piggyback from the ANGUS-003 vehicle. Two U.S. Navy programs are in operation at Naval Research Laboratories in Washington, D.C., and the Naval Ocean Systems Center for developing a vehicle to perform magnetic pipeline following. This work is in conjunction with the U.S. Geological Survey.

University of New Hampshire engineers have built an untethered vehicle to follow pipelines (from the outside) via acoustic arrays. The vehicle weighs 372 kg and operates up to 8 hours. MIT is working on a small search and survey vehicle designed for low acoustic noise. ROBOT III includes a rubberized fabric as a protective outer liner or skin and floodable tanks for automatic ballast and trim. Considerable work is being pursued on fiber optic and acoustic communication links. It is expected that "intelligent" systems will improve untethered vehicles to the point that most tethered ROVs will not be used for work at great depths.[12]

Given (1979) provides the most detailed review of remotely controlled vehicles (RCVs) and provides photographs of many of the existing systems. Most of the systems are tethered. However, one free, swimming system will be used to inspect the bottom of icebergs that may ground and destroy equipment on the sea bottom. If a danger is present, the berg can be towed away. Iceberg towing techniques have been developed and demonstrated by the University of Newfoundland.

---

[12] Marvin Minsky, at MIT, has suggested the use of robots to find and repair leaks in nuclear reactors or in deeply buried water pipes. A recent innovative application of ROVs is being studied by a French firm in a bid to produce a vehicle to inspect the network of sewers and drains under the city of Paris.

OTHER FUTURE USES OF ROBOTS

Mining coal and exploring space and the ocean floor do not exhaust the possibilities of robots. Handling dangerous radioactive wastes on a routine basis in a future disposal facility is one example.[13] The choice is between one kind of mechanization and another: human workers cannot be routinely exposed to these wastes. Mobile robots would offer a much greater degree of flexibility than teleoperators or "hard" automation.

Finally, prosthetic robots, retail robots, and household robots exemplify service categories that are increasingly needed and difficult to obtain in any other way. Paraplegics, and especially quadriplegics, for instance, might be served full-time by voice-activated robots capable of doing a variety of necessary tasks from feeding to turning pages. Such robots are being developed in Japan. In the United States, the Veterans Administration has an ongoing program in rehabilatative robotics. The all-purpose household "droid" robot is probably a rather visionary idea, at present, but robots could certainly be designed to perform some types of jobs, notably heavy cleaning. Joseph Engelberger, president of Unimation, has promised that he will soon have a robot (to be named Isaac, after Asimov) that will serve coffee in his office. Quasar Industries of Rutherford, New Jersey, built and photographed a model household android in 1978 and announced their optimistic intentions to mass produce it before 1980. The project was a hoax, but there is still unquestionable commercial interest in developing such a product, if only because of the vast potential market. In fact, Nieman-Marcus Department Stores advertised a household robot (actually a radio-controlled machine) in their 1981 catalog. For every conceivable application of an industrial robot, we can think of at least ten applications for a household robot. It is impossible to believe that such a vast market will not be exploited at the earliest possible time. In the interim, there may be a more immediate market for voice-activated robots in some retail shops and fast-food restaurants to find, display, and package goods. A robot hamburger cook, a robot clerk

---

[13] A teleoperator was designed by Hughes Aircraft in 1958 to handle radioactive materials at Atomic Energy Commission facilities in Albuquerque, NM. The U.S. Department of Energy is currently applying robotics to nuclear reactor maintenance.

in a hardware store, and a robot soldier are interesting possibilities.

It is vitally important to recognize the potential importance of some of these applications—and some of their adverse consequences—in the picture as a whole. It is entirely conceivable, for instance, that a century hence historians looking back might say, in effect, "The real significance of robotics development in the 1980s and 1990s is that they enabled mankind to expand his abode permanently beyond the earth's surface, and thereby escape the trap of limited resources associated with that constraint." All of future history could be very different, depending on whether space is successfully "colonized" in the next century or not. On the other hand, future historians might find that the household robot or the soldier robot, or even the "pet" robot, turned out to be the most important development of the next two decades.

REFERENCES

Adams, John W. The Winnig of Ore—Will Mining Ever Become a Science? In A.G. Chynoweth and W.M. Walsh, Jr. (editors), *Proceedings of the APS New York Meeting*. American Institute of Physics, New York, 1976.

Arrhenius, Gustauf. *Mineral Resources on the Ocean Floor*. Technical Report, Scripps Institute of Oceanography, 1974. Presented at the New Wave of Exploration in the Earth Sciences Symposium, Massachusetts Institute of Technology, December 3–4, 1974.

Arthur D. Little, Inc. Conceptual Design of a Fully Automated Continuous Mining System Operating under Remote Supervisory Control. Prepared for U.S. Department of the Interior, Bureau of Mines.

Babcock, William H., Steven B. Siegel, and Christina A. Swanson. *Energy Demands: 1972 to 2000*. Technical Report PRC–D–2017, PRC Energy Analysis Company, May 1978. Prepared for the Environmental and Resource Assessment Branch, Division of Solar Technology, U.S. Department of Energy.

Bender, E. Remotely operated vehicles continue rapid growth offshore. *Sea Technology:*14–19, December 1980.

Boller, J., and F.R. Busby. Ocean and Coastal Engineering Division of the Marine Technology Society: Undersea Vehicles Committee. *Marine Technology Society Journal* 14(4):4–6, 1980.

Booda, L.L. Manned Submersibles Still Hold Upper Hand. *Sea Technology*:10–12, December 1979.

Bortz, Fred. Robots: A New Breed of Blue Collar Workers. *Carnegie-Mellon University Alumni News* 64(3):3–5, September 1980.

Breslin, John, and Richard J. Anderson. *Observations on Current American, British and West German Underground Coal Mining Practices.* Battell Memorial Institute, 1976.

R. Frank Busby Associates, Inc. *Remotely Operated Vehicles.* Technical Report #03–78–G03–0136, NOAA, Contract, August 1979.

Criswell, David, et al. *Extraterrestrial Materials Processing and Construction.* Technical Report Lunar and Planetary Institute, Houston, Texas, 30 September 1978.

De Rocher, W.L., Jr., et al. *Integrated Orbital Servicing Study Follow-On.* Technical Report NAS 9–30820, Martin-Marietta Corporation, Denver Division, June 1978.

De Rocher, W.L., Jr. *Integrated Orbital Servicing Study for Low Cost Payload Programs.* Technical Report, Martin-Marietta Corporation, September, 1975. Denver Division Report to NASA Marshall Space Flight Center under Contract NAS8–10820, Volume 1 Executive Summary, Volume II Technical and Cost Analysis.

Dixon, J.K., H.A. Johnson, and J.R. Slaghle. The Prospect of an Underwater Naval Robot. *Naval Engineering Journal,* February 1980.

Frisbie, F.R. Unmanned Submersible Role Expanding. *Petroleum Engineer International* 51(6):32–40, May 1979.

Geyer, R.A. *Elsvier Oceanography Series 17: Submersibles and Their Use in Oceanography and Ocean Engineering.* Elsevier Scientific Publishing Co., New York, 1977.

Given, D. More Performance and Scope from Latest RCV's. *Offshore Services* 12(6):49–51, June-July 1979.

Gordon, G.D., and W.L. De Rocher, Jr. Repairing a Communications Satellite in Orbit. *COMSAT Technical Review* 6(1), Spring 1976.

Gordon, G.D. A User Assessment of Servicing in Geostationary Orbit. *Mechanism and Machine Theory* 12(5), October 1977.

Gordon, G.D., et al. *Integrated Orbital Servicing and Payloads Study.* Report to NASA, Marshall Space Flight Center under contract NAS 8–30849 September, COMSAT Corporation, September 1975.

Harold, Robert. Continuous Miner Manufacturers: Each Offers Something Special. *Coal Age,* February 1980.

Heer, Ewald. New Luster for Space Robots and Automation. *Astronautics and Aeronautics* 16(9):48–60, September 1978.

Heer, Ewald. Robots in Modern Industry. *Astronautics and Aeronautics*:50–59, September 1981.

Jarrett, N., S.K. Das, and W.E. Haupin. Extraction of Oxygen and Metal From Lunar Ores. *Space Solar Power Review* 1:281–287, 1980.

Johnson, H.A., A.J. Verderese, and R.J. Hanson. A "Smart" Multi-Mission Unmanned Free Swimming Submersible. *Naval Engineering Journal*, April 1976.

Korf, Richard E. *Space Robotics*. Technical Report, Carnegie-Mellon University Robotics Institute, May 1981.

Lindbergh, Kristina and Barry Provorse (editors). *Coal: A Contemporary Energy Story*. McGraw Hill, New York, 1979.

Long, J.E., and T.J. Healy (editors). *Advanced Automation For Sapce Missions, Technical Summary*. Technical Report, NASA/University of California at Santa Clara, September 1980. Summary of the NASA/U. of California at Santa Clara/ASEE summer study conducted June 23–August 29, 1980.

Lund, Robert, David R. Pinsky, and James R. Showalter. *Computer-Aided Materials Processing*. Technical Report CPA–80–3, Center for Policy Alternatives, Massachusetts Institute of Technology, July 1980.

Morley, R.A. RCV's as Precision Tools. *Underwater Technology Journal* 5(2):9–13, June 1979.

National Aeronautics and Space Administration. *A Selected Bibliography for Teleoperator and Remote Systems Technology, 1966–1982*. Technical Report, NASA Marshall Space Flight Center, 1982.

National Academy of Science. *Energy in Transition: 1985–2010*. W.H. Freeman and Co., San Francisco, 1979. Final Report of the Committee on Nuclear and Alternative Energy Systems (CONAS) of the National Research Council.

Fleet of Unmanned Submersibles Continues to Grow. *Offshore* 39(9):73–82, 1979.

Panek, L.A. and J.A. McCormick. *Roof/Rock Bolting*. Society of Mining Engineers, AIME, New York, 1973.

Pennsylvania State University. Study of the Human Factors Aspects of an Automated Continuous Mining Section. U.S. Bureau of Mines, Washington, D.C.

President's Commission on Coal. *Coal Data Book*. Government Printing Office, Washington, D.C., 1980.

Romro, James. *Advanced Space Transportation Technology Program*. Technical Report, NASA Headquarters, Washington, D.C., August 1981.

Schmidt, Richard A. *Coal in America, an Encyclopedia of Reserves, Production and Use*. McGraw Hill, Inc., New York, 1979.

Bureau of the Census, U.S. Dept. of Commerce. *Historical Statistics of the United States, Colonial Times to 1957*. Government Printing Office, Washington, D.C., 1961. A Statistical Abstract Supplement.

Bureau of the Census, U.S. Dept. of Commerce. *Statistical Abstract of the United States, 1979.* Government Printing Office, Washington, D.C., 1979. 100th ed.

Walton, Daniel R., and Peter W. Kanttman. Preliminary Analysis of the Probable Causes of Decreased Coal Mining Productivity. Prepared for U.S. DOE, Division of Solid Fuels Mining and Preparation.

Warner, E.M. Joy Pushbutton Miner. Paper L–7082 presented before the district AIEE meeting.

White, W., and M. Holmes. The Future of Commercial Satellite Telecommunications. *Datamation,* July 1978.

Yastrebov, V.S., and Stefanov, G.A. Underwater Robot/Manipulator Development. *Marine Technology Society Journal* 12(1):3–9, 1978.

5 EMPLOYMENT IMPACT OF ROBOTIZATION

The extent to which industrial robots gain acceptance in the United States in the decades ahead will be significantly influenced by two sets of considerations. One concerns engineering and technical issues such as those discussed in Chapter 2. Superior technical performance leading to greater productivity and reductions in cost, primarily in labor costs, are mainly responsible for creating the potential for more widespread use of the new technology. The other set of considerations relates to the social costs of robotization that result from the displacement of semiskilled and unskilled workers from their current jobs. Emphasis on the social costs runs the risk of suggesting possible "remedies" that would substantially slow down the introduction of robots by adopting arrangements to protect labor. Such measures would significantly increase the financial burden on hard-pressed firms in slow-growing (or declining) industries. An assessment of the consequences for employment of the more general use of robots can perhaps supply a basis for arriving at policies that minimize the human dislocations involved without introducing major obstacles to technical advances.

The limited experience to date suggests that for the overall economy, industrial robots have posed little, if any, threat to employment. It is estimated that there were 1,600 robots in use in the United States in 1977 and roughly 4,100 by the end of 1981

187

(see Chapter 1). This suggests that robots may have displaced 3,000 to 5,000 workers by 1977 and 8,000 to 12,000 by 1980. The maximum estimate of 10,000 jobs lost so far would represent about one-seventh of 1 percent of the approximately 8 million semiskilled operatives employed in manufacturing industries in the United States. From this perspective, the effects on employment up to now have been negligible. But extrapolating the experience of the recent past into the future overlooks the concentration of effects in a relatively narrow industrial sector, occupational, and regional setting. In particular, robotization— along with other and frequently related developments—could diminish employment opportunities for semiskilled operatives in durable goods industries, especially in metal-working industries. These are industries that are frequently concentrated in the already lagging East North Central and Middle Atlantic regions. These considerations imply that the displacement will be of sufficient magnitude, at least in some industries and regions, and for some groups of workers, to be a cause of concern.

Four questions arise in appraising the employment impacts of robotization. They are:

1. Would more widespread use of robots create a special employment problem—as contrasted with problems associated with automation generally—and for whom would it be a problem?
2. What are the estimates of the potential displacement likely to be created by robots over the next two decades and how reliable are the estimates?
3. Would more general use of robots have regional consequences for employment more far-reaching than its impact on the overall economy?
4. Are existing collective bargaining and labor market arrangements likely to provide remedies for the displacement created by robots while minimizing the barriers to their use?

AGGREGATE TRENDS IN MANUFACTURING EMPLOYMENT

Developments such as robots and computerized manufacturing (CAM) continue the long-term trend toward standardization, mechanization, and computerization that has made for relatively static employment in manufacturing industries in the United States, despite increased output. The tendency to substitute capital equipment for labor is illustrated by the experience in the automobile industry since the end of World War II. In 1948, some 713,000 U.S. and Canadian auto workers produced 5.96 million new cars, trucks, and buses. In 1978, 839,000 auto workers in the two nations produced nearly 14.26 million motor vehicles. Seventeen motor vehicles were produced per automobile worker in 1978 as compared with 8.36 vehicles in 1948 (United Auto Workers 1980). These figures certainly understate the growth in output per worker since they do not take into account the increase in complexity of motor vehicles in the past generation. The net effect of these changes shows up in terms of slower growth in the employment of production workers in manufacturing than in total employment (see Table 5–1).

Total nonagricultural payroll employment nearly doubled in the twenty-year period between 1948 and 1978, whereas jobs for production and nonsupervisory employees in manufacturing industries increased by less than one-seventh. More recently, during the decade of the 1970s nonagricultural employees increased by 19 million. The comparable increase for production workers in manufacturing was 1 million. Between 1979 and 1980, the total

Table 5–1. Number of Persons on Payrolls of Nonagricultrual Establishments, 1948 to 1980.

Year	Total Employees (in millions)	Index (1948 = 100)	Employment for Production Workers in Manufacturing (in millions)	Index (1948 = 100)
1948	44.9	100.0	12.9	100.0
1960	54.2	120.7	12.6	97.5
1970	70.9	157.9	14.0	108.8
1978	86.7	193.1	14.7	113.9
1979	89.9	200.2	15.1	117.0
1980	90.7	202.0	14.3	110.9

Source: Employment and Training Report of the President (1981: Tables C1-C2: 211-212).

number of employees continued to increase, but high rates of unemployment in manufacturing brought the employment of production workers down by 700,000 workers, near the 1970 level.

The beginnings of a shift to a postindustrial society are evident in several ways. The percentage of the workforce employed in producing goods (agriculture, manufacturing, and mining) has steadily declined from nearly one-third of the workforce in 1959 to barely over one-quarter in 1977. Projections indicate this trend will continue (Table 5–2).

In addition, the "service content" of each major sector of the economy, as represented by the proportion of the industry's workforce classified as "nonproduction" workers, has steadily increased. Over one-quarter of the workforce in manufacturing and mining is in managerial, professional, clerical, sales, or supervisory activities (Table 5–3). The professional category includes lawyers, accountants, scientists, engineers, computer specialists, planners, personnel specialists, and others.

Slow growth in manufacturing employment in the 1970s has contributed to the substantially higher unemployment rates in that decade than in the previous one. The economy was consequently unable to provide employment for all the women wanting to enter the labor force or the "baby boom" generation born in the late 1950s and early 1960s. The significance of robotization in the next decade or two is that it will add to the combination of factors that have retarded growth in employment in manufacturing.

Table 5–2.   Percentage of Workforce Employed in Goods and Service Sectors.

Sector	1959	1968	1977	1990 (projected)
Agriculture, mining, and manufacturing	34.4	30.5	25.2	23.2
Government, construction, utilities, transportation, trade, and services	65.6	69.5	74.8	76.8

Source: Bureau of Labor Statistics (1979a: 32)

Table 5–3.   Nonproduction Workers as Percentage of Total Employment by Major Sector.

Sector	1950	1960	1970	1980
Mining	9.4	19.9	24.1	25.6
Construction	11.1	14.7	16.7	21.3
Manufacturing	17.8	25.1	27.5	29.9
Transportation	b	11.7[a]	13.3	16.5
Trade	6.9	9.4	11.0	12.3
Finance, Insurance, Real Estate	17.1	18.4	21.0	24.4
Services	b	8.3[a]	9.2	11.0

[a] Data for 1964
[b] Not available
Source: Employment and Training Report of the President (1981: Table C-3: 213).

OCCUPATIONAL EMPLOYMENT IN MANUFACTURING

In 1980, about 14 million people were employed as production workers in manufacturing. Nearly half of all production workers (and for that matter, nearly half of all manufacturing workers) are concentrated in the metalworking industries. This sector includes the following industries, followed by their standard industrial classification (SIC) codes:[1]

- Primary metals (SIC 33)
- Fabricated metal products, except machinery and transportation equipment (SIC 34)
- Machinery, except electrical (SIC 35)
- Electrical and electronic machinery, equipment, and supplies (SIC 36)
- Transportation equipment (SIC 37)
- Precision Instruments (SIC 38)

Although no official statistics of robot use by industry are collected in the United States, it appears that at least 80 percent of

---

[1] A list and description of all detailed industries included in SIC 33-38 is given in (Census of Manufacturers 1981a, Appendix C).

existing robots are used within these industries. Official statistics from Japan confirm that about 85 percent of all robots sold there in 1978 went to the corresponding sectors (Yonemoto 1981).

Manufacturing production workers are classified into three main categories (Table 5–4): craft (skilled workers), operative (semiskilled workers), and laborers (unskilled workers). Examples of occupational titles within each of these major groupings are shown in Table 5–5. As noted in the overview of robot technology in Chapter 2, robots without sensory feedback perform successfully in simple repetitive, well-structured tasks that can be preprogrammed, such as spot welding, spray painting, palletizing, and materials handling. Sensor-based robotic systems can accom-

Table 5–4. Employment of Production Workers, 1980: Metal-working and Total Manufacturing.

Occupation	Employ- ment in Metalwork- ing SIC (33-38)	Total Employ- ment, All Metalwork- ing	Percentage Employ- ment in Manufac- turing
Total, all occupations	9,964,878	20,361,568	48.9
Production workers, total: (craft workers, operatives, laborers)	6,688,306	14,190,289	47.1
Craft and related workers, total	2,015,212	3,768,395	53.4
Metalworking craft workers	582,861	668,002	87.2
Other craft workers	1,432,351	3,100,393	46.1
Operatives, total	4,060,916	8,845,318	45.9
Nontransport operatives	3,880,876	8,134,123	47.7
Assemblers	1,311,870	1,661,150	78.9
Metalworking machine operatives	1,030,132	1,069,540	96.3
All other machine operatives	893,701	4,231,988	21.1
Welders and flame cutters	369,558	400,629	92.2
Production painters	79,594	106,178	74.9
Packing and inspection operatives	78,413	587,631	13.3
Sawyers	17,604	76,728	22.9
Transport operatives	180,040	711,195	25.4
Laborers, except farm	612,178	1,576,576	38.8
Nontransport operatives and laborers	4,493,054	9,710,699	46.2

Source: Compiled from Bureau of Labor Statistics (1982a).

Table 5–5.   Representative Occupational Titles for Craft Workers, Operatives, and Laborers.

Metalworking craft workers, except mechanics	Other craftworkers
Blacksmiths	Electricians
Core makers	Plumbers
Forging press operators	Pipefitters
Heat treaters, annealers, temperers	Mechanics and installers
Layout makers, metal	Blue-collar supervisors
Machinists	Heavy equipment operators
Metalworking machine tool setters	
Metal molders	
Punch press setters	
Sheet metal workers and tinsmiths	
Tool and die makers	
Metalworking machine operatives	**Other machine operatives**
Casters	Machine operators for textile products
Drill press and boring machine operators	Machine operative for wood products
Electroplaters	Machine operators for rubber products and plastics
Furnace charges and operators	Machine operators for food products
Grinding, abrading machine operators	Machine operators for paper products
Lathe machine operators	Machine operators for chemical products
Machine tool operators, combination	
Machine tool operators, numerically controlled	
Machine tool operators, tool room	
	Laborers
	Conveyor operators and tenders
Milling and planning machine operators	Furnace operator helpers
	Off-bearers
Metal punch press machine operators	Loaders, car and truck
	Stock handlers
Power brake and bending machine operators	
Punch press operators	

For a more complete list of occupational titles within each group, see Bureau of Labor Statistics (1982a).

plish tasks involving a greater degree of variability and to a limited (but rapidly increasing) degree can select workpieces regardless of the order in which they arrive, assemble small components, and carry out dimensional inspections.

The occupational titles show that most of the routine repetitive jobs that currently lend themselves to automation and robotization are performed by the semiskilled (nontransport) operatives who comprise nearly 40 percent of total manufacturing employment. The nontransport operatives are an important factor in only one other sector—mining—where they made up 29 percent of the employed workforce in 1979. Thus mining appears to be a plausible future application for robotics. In the past generation operative jobs in manufacturing have provided employment and economic and social mobility to blacks migrating from the rural South to the North, to poor whites seeking to escape poverty in Appalachia as, earlier, they provided entry-level positions to immigrants. The more than 8 million manufacturing operatives in 1980 included a greater than proportionate share of blacks and other nonwhites than in the overall civilian labor force (Figure 5–1). Nonwhites account for only 11 percent of the national workforce but comprise over 16 percent of total employment in semiskilled and unskilled manufacturing jobs. Blacks and other nonwhites were 53 percent more likely to be employed in these occupational areas than whites.[2] Nontransport operatives (mostly semiskilled and unskilled manufacturing workers) have traditionally been the largest single group of unemployed workers. The unemployment rate for this group jumped sharply from 8.4 percent in 1979 to 12.2 percent in 1980, whereas the overall unemployment rate increased somewhat less preciptiously from 5.8 percent to 7.1 percent over the same one-year period. Correspondingly, the prospect of a nontransport operative becoming unemployed in 1980 exceeded the overall probability for the labor force as a whole by 72 percent. The average unemployment rate for non-

---

[2] This assertion is based on the following figures, derived from the employment statistics in Table 5–1:

Proportion of whites employed as semiskilled and unskilled manufacturing workers: 7,412,922/86,375,760 = 8.6 percent.

Proportion of nonwhites employed as semiskilled and unskilled manufacturing workers: 1,435,078/10,894,240 = 13.2 percent.

13.2 percent/8.6 percent = 1.53 percent.

Figure 5–1. Distribution on the Manufacturing Workforce by Sex and Race, 1980.

Percentage distributions:

M: male

F: female

W: white

NW: non white

Total Employed Persons: 97,270,000

	W	NW	Totals
M	51.7	5.8	57.5
F	37.1	5.4	42.5
Totals	88.8	11.2	100.0

SKILLED WORKERS

Machine Jobsetters: 658,000

	W	NW	Totals
M	88.2	7.9	96.1
F	3.2	.7	3.9
Totals	91.4	8.6	100.0

Other Metalworking Craftworkers: 638,000

	W	NW	Totals
M	89.2	6.9	96.1
F	3.3	.6	3.9
Totals	92.5	7.5	100.0

SEMI SKILLED AND UNSKILLED WORKERS

Moter Vehicle Equipment Operatives: 431,000

	W	NW	Totals
M	65.7	14.6	80.3
F	15.1	4.6	19.7
Totals	80.8	19.2	100.0

Other Durable Goods Mfg. Operatives: 4,166,000

	W	NW	Totals
M	55.7	8.5	64.2
F	30.2	5.6	35.8
Totals	85.9	14.1	100.0

Non Durable Goods Mfg. Operatives; 3,290,000

	W	NW	Totals
M	35.0	6.6	41.6
F	47.3	11.1	58.4
Totals	82.3	17.7	100.0

Manufacturing Laborers: 961,000

	N	NW	Totals
M	68.0	16.3	84.3
F	13.0	2.7	15.7
Totals	81.0	19.0	100.0

Source: Derived from unpublished data supplied by the Bureau of Labor Statistics, Employment Structure and Trends Division.

whites in 1980 was nearly 2.1 times higher than the average for whites. There may also be a disproportionate impact on women, who account for nearly 41 percent of unskilled and semiskilled manufacturing jobs (3.6 million). Women in these jobs are much less likely to be represented by labor organizations than their male counterparts.[3] It seems that a major expansion of robotization and related advances in computer-controlled manufacturing processes may have a disproportionate impact in creating job losses among groups that are already substantially disadvantaged in the nation's labor markets.

Metalworking employment

Nearly half of all nontransport operatives are employed in the metalworking industries (SIC 33-38). On the whole, this population is representative of all production workers. In several job categories that are most amenable to robotization, however, almost all of the employment is concentrated in metalworking. For instance (not surprisingly), nearly all the 1 million operatives of metalcutting and metalforming machines are employed in SIC 33-38. In addition, almost all assemblers, welders, flame cutters, and production painters are employed in these industries. While the percentage of packing and inspection operatives, sawyers, and laborers is comparatively small, it is probably true that most early robot applications in these jobs will be in metalworking since metal products are most suitable for robot handling.

While the majority of jobs that can be robotized are semiskilled operative jobs, there are already robot applications in heat treating, sheet metal work, and forge and hammer operations, all of which are classed as skilled jobs. As computer-assisted design (CAD) and CAM become more integrated, and factories are redesigned to exploit robotics and other types of programmable automation fully, a larger fraction of the skilled metalworking craft will be within the domain of jobs which could be automated. An approximate upper bound of the number of manufacturing jobs which could—in principle—be robotized (or integrated with various types of automation) within the next several decades is given

---

[3] See (Bureau of Labor Statistics 1981a) for the sex/race distribution of manufacturing workers represented by labor organizations.

by the number of nontransport operatives and laborers. There are 5.3 million such workers in SIC 33-38, and 11 million such workers throughout all manufacturing sectors.

ESTIMATING POTENTIAL DISPLACEMENT

Estimates of the employment displacement created by industrial robots ten, twenty, or thirty years from now are "what if" projections based on scenarios of future technological, economic, and social changes. Such projections depend on assumptions about future growth in gross national product (GNP) and about technical improvements that will increase the range of operations robots can perform that are not yet in common use or are still on the drawing boards. The techniques normally used in generating employment projections are ill-suited to anticipating the long-term impact of quantum changes in the economic environment such as the changes implied by the widespread use of robotic and CAM technologies. Allowing for these caveats, our projections represent an effort to quantify the importance of developments which evidence suggests will have significant employment impacts in the future.

Experience with automation over the past two decades underscores the tentative nature of any such estimates. Previous study groups such as the Ad Hoc Committee on the Triple Revolution, sponsored by the U.S. Congress, contended in 1964 that the employment displacement from automation would be so great that it would be necessary for the federal government to provide generous and costly income supports to a large fraction of the workforce, in order to permit them to continue as consumers in the new society in which their services would no longer be required as producers. As employment continued to grow rapidly in the 1960s, especially in service industries, the forecasts of discontinuous declines of employment resulting from technological advances lost credibility generally in the United States. In the fifteen-year period between 1964 and 1979, total civilian employment was to confound the Ad Hoc Committee's forecast by increasing by 27.6 million, or 40 percent.

The automation controversy underscores, among other things, that what is technologically possible may not be economically feasible. New scientific and engineering ideas do not automatically lead to innovations in the economic system unless there is a market for them. In the case of robots, there was no significant market from 1959 to 1979. The contention that there will be a significant expansion in the use of robots is based on many signs that an industrial market now exists for this technology and that it is growing. Indicators of the magnitude of growth that can be expected were reviewed in Chapter 1.

Wide variations in estimates of the future use of robots underscore the margin of uncertainty surrounding projections of their future impact for employment. For instance, a research group at the Illinois Institute of Technology estimated that 17,000 robots will be in use by 1985 and 70,000 by 1990. Other projections such as one by Bache Halsey Stuart Shields, Inc., anticipate that as many as 200,000 robots will be in use by 1990 (Conigliaro 1981). This projection was predicated on the probability—now confirmed as fact—that large computer and electrical equipment manufacturers such as General Electric and IBM would enter the field and produce robots on a sufficiently large scale to reduce purchase prices to users sharply.

The extent of job displacement in a given industry depends on many factors: the current level of automation within the industry, the future development of robot capabilities, the occupational structure of the industry in question, the expected cost savings from robotization, and future economic growth, not to mention retirement and turnover rates, union contracts, and government policies affecting technology change. Within the metalworking industries, there are large differences in wage rates, in the distribution of occupational employment, in the type and degree of automation already in place, and in the distribution of small and large establishments. These differences are likely to be important in determining the rate at which robotics are introduced into specific establishments, and the impact of their use on the workforce and workplace. A detailed analysis of robot/worker substitutability by industry would involve a breakdown of each job into component tasks, a description of how a human worker currently carries out each task (and how it should most rationally be carried out), and finally an analysis of the potential for robotic substitution based on technical capabilities. A general specification of task for human factory workers, of capabilities or skills required to perform each task, and of the potential for robotic substitution is shown in Table 5–6.

An economy-wide analysis for all operative jobs at this level of detail is out of the question at present, but the direction the analysis should take is apparent. Each job can be characterized by the skills required to perform the component tasks. A classification system devised for this purpose is available in the *Dictionary of Occupational Titles* (DOT), published by the Employment and Training Administration (1977) of the Department of Labor.[4] Part of the DOT skill classification system is shown in Table 5–7.

Supplementary information in the DOT classification scheme is shown in Table 5–25 at the end of this chapter. Unfortunately, the DOT skill classification system is not sufficient for identifying whether a particular task could be done with a robot, since it does not give detailed enough information on the dexterity and sensing requirements for each task. Another approach to determining whether a robot could be substituted for a worker is to carry out a detailed engineering analysis of the basic tasks required to perform a particular job based on time and motion studies of both humans and robots (Nof and Lechtman 1982). However, this would require modelling every different manufacturing job, a near impossible task in itself. New skills analysis tools are needed by labor planners to predict whether robots and other forms of automation can be used to perform specific manufacturing tasks, without explicitly modelling evey job in detail.[5]

Even within a single occupational title, such as "manufacturing inspector," there is a wide range of skill and physical requirements that depend on the physical properties of the product and the production environment. The percentage of jobs within a specific occupation that could be robotized depends on how the various tasks are distributed over the range of physical and skill requirements. An example of the distribution of skill requirements for an inspector is shown in Table 5–8. An illustrative list of the distribution of physical skill requirements for several other operatives is presented in Table 5–24 at the end of this chapter.

Estimates of the percentage of workers that *potentially*could be replaced by robot technologies can be developed from an analysis

---

[4] The Dictionary of Occupational Titles and its skill classification system is the only available comphehensive guide to the large number of specific occupational titles and to skill requirements.

[5] We are in the process of trying to devise such tools.

Table 5–6.   Analysis of the Potential for Robotic Substitution by Task.

FUNCTION	HUMAN OPERATIVE		ROBOT OPERATIVE	
Select workpiece	H1	Uses senses of sight and touch.	R1	Visual methods are prohibitively difficult at the present state of the art. The practicable methods are restricted to sensing physical contact, and by a pre-programmed command which directs the robot to the workpiece location.
Picking up	H2	Uses combination of arms, hands and body. May need mechanical assistance with heavy lifts.	R2	Strictly, the robot only needs one arm and a hand, although these must be jointed to allow the hand to move in three dimensions and to swivel. The use of hydraulic power can produce greater lifting capacity than a man's.
Placing	H3	Similar to H2, but uses his eyes to ensure that the workpiece is correctly orientated.	R3	Similar to R2, but without the sense of sight, orientation becomes difficult. A robot operator needs workpieces that are presented to it in some pre-defined and constant attitude.
Operate machine	H4	Uses any one or all of his five senses to follow the operation of the machine and activate the controls as necessary. Has a memory, with which he can learn the sequence and timing of operations.	R4	Because it is not able to see, hear, or otherwise witness the progress of the machine, a robot must be pre-programmed to carry out its operations according to a timed sequence. A man has to do the teaching, and the robot has to have an internal memory to store the information. Computer technology has made this possible.
Unload machine	H5	This is similar to H2.	R5	Similar to R2.
Stack the workpiece	H6	This is similar to H3.	R6	Similar to R3, but orientation is less of a problem, because the machine will usually be one which holds every workpiece in the same position and attitude.
Inspect the workpiece	H7	A man should be capable of inspecting, as required, using his senses of sight and touch, or by using gauges and other measuring equipment. He is able to speak, and to tell the foreman of any machine problem.	R7	A robot would be capable of inspection by automatic gauging, by probing, and by telemetry or datalogging. Although not able to speak, the robot could be designed to give an audible or other warning of any defect in the workpiece.

Machine breakdown	H8	Able to use any of his senses to perceive possible-trouble. For example, he could recognize the smell from a hot motor, and switch off the machine.	R8	A robot would be capable of detecting a machine breakdown only where this produce defects in the workpieces, that the robot could detect, or where telemetry points were set up in the machine.
Machine setting	H9	This operative is unskilled or semi-skilled. He must, therefore, expect to have to call for skilled assistance whenever the machine needs resetting or retooling.	R9	The robot could be programmed to stop after a period calculated to allow for tool wear. It is also practicable to command the robot to probe the workpiece and/or the machine for broken drills, taps, reamers and other tools. A skilled human operator is needed to carry out rectification and resetting.
Shift periods	H10	A man has to leave for home after a shift period of between eight and twelve hours. He has to be mobile—able to walk and run.	R10	There is no need for a robot to be self-mobile. It can be fixed to the floor, and commanded to work round the clock every day. Where three shifts are operated within 24 hours on one machine, a robot operative can take the place of three men—more men if the robot can be set up between two or more adjacent machines that it can operate in sequence.
Workbreaks	H11	A man has to take time off work for eating, and for other physical needs.	R11	Robots do not need time off for eating or for other natural functions. They will be subject to down time for maintenance and, however reliable the design, there will inevitably be occasional breakdowns.
Strikes, go-slows and overtime bans	H12	Collective disruptive action from the labor force is a problem that can hit any production plant.	R12	A robot will obey commands slavishly and there is no labor organization for robots.
Illness and fatigue	H13	Men get tired, sometimes feel below par, and they may be absent from the workplace when they fall sick. These conditions can be aggravated or caused by unpleasant working accomodation (heat, noise, toxic fumes, smells, cramped space, etc).	R13	Robots can be expected to break down very occasionally (see R11). But it is possible to design robots that can operate in extreme conditions which might damage a man's mental or physical health—such as noise, heat, vibration, noxious fumes, smells, cramped work space and even radioactivity.

Source: Engelberger (1980: 6).

Table 5–7.   Skill Classification Used in the *Dictionary of Occupational Titles.*)

A job's relationship to *data, people,* and *things* can be expressed in terms of the lowest numbered function in each sequence. These functions taken together indicate the total level of complexity at which the worker performs. The digits express a job's relationship to data, people and things by identifying the highest appropriate function in each listing.

Data	People	Things
0 Synthesizing	0 Mentoring	0 Setting up
1 Coordinating	1 Negotiating	1 Precision working
2 Analyzing	2 Instructing	2 Operating, controlling
3 Compiling	3 Supervising	3 Driving, operating
4 Computing	4 Diverting	4 Manipulating
5 Copying	5 Persuading	5 Tending
6 Comparing	6 Speaking, signaling	6 Feeding, offbearing
	7 Serving	7 Handling
		8 Taking instructions Helping

Definitions of levels 0-8 for a job's relationship to *Things* are given in Table 5–23.

Source: Employment and Training Administration (1977: 1369-71).

of industry employment by occupation, by skill level, and from industry estimates of the potential for robotization. As part of the previously mentioned Carnegie-Mellon (1981) study, members of the Robot Institute of America were surveyed to gather data on the impacts of robotics on the workplace and to estimate the potential for robotization within various occupations. The respondents were either experienced robot users or were in the process of evaluating robot applications. The pooled sample of questionnaire respondents and those interviewed by phone accounted for nearly one-third of the U.S. robot population as of January 1981. The RIA members were asked to estimate what percentage of jobs within a given occupation could be done by a robot similar to those on the market in 1981 (Level I) and by the next generation of robots with rudimentary sensory capabilities (Level II). In total, 16 firms fully or partially completed this part of the survey.

Table 5–8. Breakdown of the Work Requirements for Inspectors.

Skill Level[a]	Job Skill Requirements	Physical Demands	Environmental Conditions	Specific Vocational Preparation
0	—	—	—	—
1	Reasoning ability	Max. lift of 25 lbs. reaching, handling, or feeling, hearing, seeing	Inside (protected from weather conditions)	> 1 year
	Sense of hearing			< 2 years
2	—	—	—	—
3	—	—	—	—
4	Visual inspection, determine which component of a system is the source of the defect	Max. lift of 50 lbs. reaching, handling, feeling, seeing	Inside (protected from weather)	< 1 year
5	—	—	—	—
6	—	—	—	—
7	Verify dimensions, visual inspection	Max. lift of 25 lbs. reaching, feeling, seeing	Inside (protected from weather	> 30 days < 6 months

[a] The job's relationship to *things*. See Table 5-7.
Source: Carnegie-Mellon University (1981: 94).

Based on the original responses, and on our own best judgment of occupational similarities, we have extrapolated the results to include occupations that were not part of the original survey. The results are shown in Table 5–9. Job setters and other metalworking craft workers are included in the table because advances in robotics, numerically controlled machine tools, and related advances in computer-assisted design and manufacturing are likely to make their jobs, like those of many operatives, susceptible to robotization. The estimates are broken down by occupation and grouped by type of machine used. Halving and doubling the estimates of the number of jobs that could be robotized by level I and level II robots roughly brackets the average response within the low and high estimates. The employment totals listed refer to the year 1977.

The 3.6 million people accounted for in Table 5–9 represent only 60 percent of all production workers in metalworking and three-quarters of the metalworking crafts, nontransport operatives, and laborers. The coverage is incomplete because some occupational titles are not included. Unclassified workers and unskilled laborers are not even listed. It appears that nearly 580,000 (or 16 percent) of the workers listed could potentially be replaced by level I robots. The figure more than doubles to nearly 1.5 million (or 40 percent) if level II robots with rudimentary sensing capabilities were available. As of 1980, there were roughly 5 million metalworking craft workers, nontransport operatives, and laborers in metalworking industries, and 10.4 million throughout all manufacturing (see Table 5–4). We assume that level I robots could theoretically replace 16 percent of the workers in these three groups and that level II robots could theoretically replace 40 percent of the same population of workers. Thus, if all the potential for job displacement of level I robots had actually been realized in metalworking in 1980, more than 800,000 jobs would have been eliminated. Assuming level II robots were available and fully exploited, an additional 1.2 million jobs, or a total of nearly 2 million jobs, could theoretically have been lost.[6] Extrapolating the data for metalworking to similar tasks in other manufacturing sectors, it appears that level I robots could theoretically replace about 1.5 million metalworking craft workers, nontransport operatives, and laborers, and level II robots could

---

[6] Two million jobs is approximately 2 percent of the total civilian work force. Kalmbach et al (1982: 43) recently report that nearly 2 percent of all jobs in the German metalworking industries could be effected by the possbile introduction of industrial robots into processing operations.

theoretically replace about 4 million out of the current total of 10.4 million of these workers. In the course of the coming decade, the capabilities of robots can be expected to increase further, and with these changes their potential for displacing operatives will increase. On the other hand, not all of the potential displacement would actually be realized by 1990, and if the economy grows, as anticipated, some of the job loss due to displacement of people by robots could be offset by increases in manufacturing employment.

PROJECTED EMPLOYMENT GROWTH IN MANUFACTURING, 1980-90

Since robotization is expected to be concentrated in the metal-working sector, projected employment trends in these industries provide a point of departure for estimating the potential displacement of employees that could take place in the 1980s. Projections of industry employment for 1990 made by the Bureau of Labor Statistics (BLS), derived from a macroeconomic model of the economy as a whole, are shown in Table 5–10). Three growth scenarios, the low trend, the high I, and the high II alternatives, are specified. The basic characteristics of the Bureau of Labor of Statistics scenarios are described by Saunders (1981: 18):

> The low-trend projection is characterized by assumptions of continuing high inflation, low productivity growth, and moderate expansion in real production. Alternatively, the high-trend version-I projection assumes marked improvements in both inflation and productivity, greater labor force growth, and commensurately higher real production levels. Finally, the new high-trend version-II alternative assumes labor force growth consistent with the low-trend, but greater productivity gains and less inflation than in the version-I high trend. None of the alternatives represents an attempt to forecast possible cyclical fluctuations during the 1980s. The three projections are intended to provide a range within which economic growth will most likely occur; however, they should not be interpreted as being representative of all likelihoods.

The assumed per annum GNP growth rate for the 1980-85 period is 2.2 percent for the low growth trend and 3.8 percent for the high I trend. In the 1985-90 period, the average GNP growth rate is assumed to increase to 2.8 percent for the low trend and 4.0

Table 5–9. Potential for Robotization.

| | Level I Robot | | Level II | | Operatives* | | Potential Displacement | | | |
	Range %	Av. %	Range %	Av. %	Sector 34-37	Sector 33-38	Sector 34-37 I	Sector 34-37 II	Sector 33-38 I	Sector 33-38 II
Drill press/boring mach.	25-50	30	60-75	65	104,050	113,210	31,215	68,933	33,963	73,587
Filer, grinder, buffer	5-35	20	5-75	35	77,360	103,430	15,472	27,076	20,686	36,201
Gearcutting, grinding, shaping		10		50	11,070	11,670	1,107	5,535	1,167	5,835
Grinding/abrading mach. op.	10-20	18	20-100	50	97,090	109,680	17,476	48,545	19,742	54,740
Lathe/turning mach. op.	10-20	18	40-60	50	130,260	141,560	23,447	65,130	25,481	70,780
Machine tool op., comb.	10-30	15	5-60	30	142,750	154,220	21,413	42,825	23,133	46,266
Machine tool op., N.C.	10-90	20	30-90	49	41,900	45,020	8,380	20,531	9,004	22,060
Machine tool op., toolroom	1-5	3	4-60	50	33,410	36,160	1,002	16,705	1,085	18,080
Machine tool op., setter		10		50	47,260	51,490	4,726	23,630	5,149	25,745
Milling/Planning mach. op.	10-20	18	40-60	50	58,900	63,230	10,602	29,450	11,381	31,614
Sawyer, metal		20		50	10,660	15,180	2,132	5,330	3,036	7,590
Subtotal, Metalcutting Machines					**754,710**	**844,850**	**136,972**	**353,690**	**153,827**	**392,599**
Coil winding	15-40	24	15-50	40	26,570	33,550	6,377	10,628	8,052	13,420
Drop hammer op.		15		70	2,990	2,990	449	2,093	449	2,093
Forging press op.		15		70	6,500	7,190	975	4,550	1,079	5,033

Occupation										
Forging/straightening roll op.	15			70	1,000	2,840	150	700	426	1,988
Header op.	20			70	5,080	5,080	1,016	3,556	1,016	3,556
Power brake/bending mach.	20			70	33,240	35,240	6,648	23,268	7,004	24,514
Press operator/plate print	20			70	4,230	4,230	846	2,961	846	2,961
Punch press op.	15	10-100	60-80	70	159,890	171,710	23,984	111,923	25,757	120,197
Punch press setter	15			70	16,080	16,840	2,412	11,256	2,526	11,788
Riveter (light)	15	5-100	10-100	30	9,090	9,090	1,364	2,727	1,364	2,727
Roll forming mach.	20			70	4,320	11,030	864	3,024	2,206	7,721
Shearer/sitter op.	20			70	22,450	28,660	4,490	15,715	5,732	20,062
Subtotal, Metalforming Machines					**291,440**	**344,310**	**49,575**	**192,401**	**56,467**	**216,060**
Conveyor op/tender	10	5-15		30	18,070	20,240	1,807	5,421	2,024	6,072
Die casting machine op.	5		10-20	10	6,530	14,670	327	653	734	1,467
Dip plater	40	20-100	50-100	77	7,780	9,500	3,112	5,991	3,800	7,315
Electroplater	20	5-40	5-60	55	27,350	29,770	5,470	15,043	7,954	16,374
Platter helper	30			100	26,100	26,560	7,830	26,100	7,968	26,560
Fabricator, metal	10			30	5,910	5,910	591	1,773	591	1,773
Fabricator, plastic	10			30	1,970	1,970	197	591	197	591
Furnace op./cupola tender	20			50	4,420	14,490	884	2,210	2,898	7,245
Heater, metal	20			100	2,070	5,010	414	2,070	1,002	5,010
Heat treater, annealer	10	5-50	5-90	46	14,770	23,440	1,477	6,794	2,344	10,782
Injection/compression mold machine op. (plastic)	20		5-60	50	24,910	29,830	4,982	12,455	5,966	14,915
Inspector	13	5-25		35	228,530	269,650	29,709	79,986	35,055	94,378

Table 5–9. continued.

| | Level I Robot | | Level II | | Operatives* | | Potential Displacement | | | |
| | | | | | | | Sector 34-37 | | Sector 33-38 | |
	Range %	Av. %	Range %	Av. %	Sector 34-37	Sector 33-38	I	II	I	II
Laminator, preforms		20		50	10,160	10,160	2,032	5,080	2,032	5,080
Machine op. NEC	10-50	16	20-65	25	13,020	38,590	2,083	3,255	6,173	9,648
Molder, machine		20		50	5,650	18,540	1,130	2,825	3,708	9,270
Packager, production	1-40	16	2-70	41	55,480	75,640	8,877	22,747	12,102	30,939
Painter, production	30-100	44	50-100	66	74,380	78,540	32,737	49,091	34,558	51,836
Pourer metal	5-20	10	10-30	24	1,280	13,280	128	307	1,328	3,187
Sandblaster, shortblaster	10-100	35	10-100	35	6,290	10,030	2,202	2,202	3,511	3,511
Screwdriver op. (power)		10		50	3,420	3,420	342	1,710	342	1,710
Tester	1-10	8	5-30	12	51,470	652,890	4,118	6,176	5,031	7,589
Wirer, electric	0-10	9	10-50	28	22,940	26,520	2,065	6,423	2,387	7,426
Subtotal, Miscellaneous Machines					**612,500**	**788,850**	**112,504**	**258,903**	**141,700**	**322,647**
Joining (Welding)	10-60	27	10-90	49	319,040	344,280	86,141	156,330	92,956	168,697
Assembly	3-20	10	20-50	30	1,182,650	1,318,750	118,265	354,795	131,875	395,625
Total of subtotals + Joining + Assembly					**3,160,340**	**3,640,840**	**503,457**	**1,316,119**	**576,827**	**1,495,628**

Source: Ayres and Miller (1982: 55).

Table 5–10. Employment in Manufacturing and Metalworking Industries, 1980 and Projected 1990 (Bureau of Labor Statistics Scenarios.)

	Employment, (000s) 1980	Projected Employment 1990 (000s)		
		Low	High II	High I
Total employment[a]	104,120.0	121,971.0	123,958.0	130,665.0
		(17.4)	(19.0)	(25.5)
Manufacturing, total	20,361.6	23,331.6	23,675.0	25,256.5
SIC 20-39		(14.6)	(16.3)	(24.0)
White-collar	6,171.3	7,308.7	7,431.1	7,943.5
		(18.4)	(20.4)	(28.7)
Craft	3,768.4	4,304.5	4,374.0	4,656.5
		(14.2)	(16.1)	(23.6)
Operatives	8,845.3	10,015.3	10,148.6	10,848.2
		(13.2)	(14.7)	(22.6)
Laborers	1,576.6	1,703.1	1,721.2	1,808.3
		(8.2)	(9.2)	(14.7)
Metalworking,	9,964.9	11,859.5	12,101.6	13,131.5
SIC 33-38		(19.0)	(21.4)	(31.8)
White-collar	3,276.1	4,025.7	4,103.8	4,460.8
		(22.3)	(25.3)	(36.2)
Craft	2,015.2	2,374.6	2,423.2	2,634.1
		(17.8)	(20.2)	(30.2)
Operatives	4,060.9	4,780.6	4,883.8	5,308.6
		(17.7)	(20.3)	(30.7)
Laborers	612.7	678.5	690.9	738.0
		(10.7)	(12.8)	(20.5)

[a] Total employment is for the year 1979 and includes total private employment plus total government employment, including military.
Numbers in parentheses indicate percentage increase from 1980.
Sources: Personick (1981) and Bureau of Labor Statistics (1982).

percent for the high I trend. The Bureau of Labor Statistics compares the real rate of growth assumed in the low growth trend to growth rates experienced in the 1973-80 period, and growth rates assumed in the two high trends to those experienced in the 1955-58 era. The five underlying major groups of assumptions—fiscal, demographic, productivity, unemployment, and prices—are discussed in more detail in Saunders (1981).

The productivity and employment estimates assume continuing technological change but make no special assumptions about the impact of robots or related technical advances for employment in the 1980s.[7] The projections are predicated on the assumption that changes in technology and productivity will be gradual, or that they can be anticipated.[8]

Similarly, the estimates assume a stable socioeconomic environment in which large changes in the structure of wages and prices, wars, or changes in priorities and lifestyles, are insufficient to have a significant effect on the economy. This version of the future is reflected in the fixed, or gradual, changes built into the industry output-employment coefficients that provide the basis for the estimates. When major changes occur, as the sudden change in oil prices due to OPEC in the 1970s, earlier projections quickly become outmoded.

The BLS projections anticipate continued growth in employment in the manufacturing sector, but at a slower rate of increase than the rate of growth in total private employment (Personick 1981). Within manufacturing, employment growth rates for for the durable goods industries (comprised mostly of the metalworking industries) are higher than growth rates for nondurable goods industries. According to these projections, by 1990 over half of all employment in manufacturing will be in the metalworking industries (SIC 33-38). Machinery, except electrical (SIC 35) and fabricated metal products (SIC 34), is projected to have the highest growth rates in employment among the six metalworking industries. At the more detailed industry level, typewriters and office equipment, and computers and peripheral equipment, both in SIC 35, are projected to be among the most rapidly growing industries throughout the entire economy (Table 5–21). On the basis the BLS projections, we see that the manufacturing industries most likely to be subject to robotization in the 1980s are also those in which above average increases in employment would otherwise be anticipated.

Our estimates of potential for robotization and of potential displacement and the above BLS projections of employment growth

---

[7] See Bureau of Labor Statistics for an overview of a recent analysis of technology, productivity, and employment trends within several metalworking industries.

[8] For an analysis of the projections methodology of the Bureau of Labor Statistics, see (Lecht 1979).

provide a basis for analyzing net job displacement in metalworking. The results of our survey and of the BLS employment projections for blue collar workers are summarized in Table 5–11. The occupational categories in Table 5–11 are defined more broadly than metalworking craft workers and nontransport operatives, so the totals for blue collar workers are somewhat larger than the totals mentioned above. Assuming level I robots were fully implemented in the metalworking industries by 1990, nearly 800,000

Table 5–11. Projected Employment Growth Versus Potential Robotic Replacement.

	Employ- ment, 1980 (000s)	Incremental Employment Projected for 1990 (000s)		
		Low	High II	High I
Metalworking: SIC 33-38				
Operatives and laborers[a]	4,673.6	+785.5	+901.5	+1,373.1
Craft workers[a]	2,015.2	+359.4	+408.0	+608.9
Blue-collar, total	6,688.8	+1,144.9	+1,309.0	+1,982.0
Estimated Potential for Robotization				
level I	800.0			
level II	2,000.0			
Manufacturing, total: SIC 20-39				
Operatives and laborers[a]	10,421.9	+1,296.5	+1,447.9	+2,234.7
Craft workers	3,768.4[a]	+536.1	+605.6	+888.1
Blue-collar, total	14,190.3	+1,832.6	+2,053.6	+3,122.7
Estimated Potential for Robotization				
level I	1,500.0			
level II	4,000.0			

Sources: 1980 employment and projections of additional employment in 1990: Derived from data in Table 5–10.

Potential for robotization: Derived from survey results shown in Table 5–9.

[a]Operatives include transport operatives. Craft workers include both metalworking craft workers and other craft workers. See Tables 5-4 and 5-5.

workers could be displaced. This number roughly corresponds to the BLS projections of the number of job openings among operatives and laborers in the same industries. Assuming level II robots were fully implemented by in metalworking by 1990, nearly 2 million jobs could be displaced. This number exceeds the projected increase for all blue collar employment in these industries. For all manufacturing, the 1.5 million workers who could potentially be displaced if level I robots were fully implemented exceeds the number of job openings for operatives and laborers under the low and medium (high-II) growth projections. A loss of 4 million jobs would exceed the total employment increases projected for all blue collar manufacturing in the high-growth projection by nearly a million jobs. These figures suggest that if robots are fully implemented over the next decade, increases in manufacturing output will not necessarily be accompanied by increases in employment requirements, especially for operatives and laborers. The extent to which the potential impacts will become translated into actual displacement of people from jobs will also depend on the rate of investment in industry, wage trends, and robot price trends and performance.

Job losses that may have a minor effect on total employment can have a magnified impact if they are concentrated in specific industries. A total loss of a million jobs would have a modest effect for total employment if they were evenly dispersed throughout the economy. The loss would amount to less than 1 percent of the total private employment level, projected to be 104 million in 1990 (low-growth trend). The equivalent job loss in the metalworking industries alone would wipe out almost all the expected increase in operative and laborer employment in the 1980 to 1990 period (see Table 5–22). The problem created by the displacement would be greater if a majority of the job losses took place in two or three years during this period and if, as expected, they were concentrated in semiskilled occupations in areas that are experiencing slow economic growth.

REGIONAL IMPLICATIONS

The employment displacement problems anticipated from robotization are likely to be compounded by regional differences in the importance of manufacturing and in employment growth. The losses in employment would primarily take place in the East North Central and Middle Atlantic states. These are the areas in which the metalworking industries—those primarily affected by robotization—are concentrated. They are also areas that have been characterized by slow growth in employment and out-migration of their younger and better educated population in recent decades. The gains in employment, especially in service industries, would be more widely dispersed around the nation.

The geographic divisions that have accounted for the highest proportions of value added in manufacturing have also been characterized by below-average employment growth in the past ten or fifteen years. The divisions with the more rapid increases in employment have been the ones in which the value added in manufacturing has been relatively minor (see Table 5–12). The East

Table 5–12. Value added and Employment Growth in Manufacturing by Geographic Division.

Division	Percentage of Value Added in Manufacturing, 1977	Percentage Change in Manufacturing Employment, 1967-79
All regions	100.0	+ 8.5
Northeast region	23.7	
New England division	6.1	- 2.8
Middle Atlantic division	17.6	-15.4
North Central region	34.3	
East North Central division	27.4	- 0.3
West North Central division	7.0	+19.6
South region	27.4	
South Atlantic division	12.4	+22.2
East South Central division	6.1	+31.3
West South Central division	8.9	+51.1
West region	14.5	
Mountain division	2.3	+77.8
Pacific division	12.2	+25.4

Sources: Value added in manufacturing: Census of Manufacturers (1981a: Table F). Employment by region: Statistical Abstract (1980: Tables 691 and 1445).

North Central and Middle Atlantic states accounted for close to half, 44 percent, of the value added by manufacturing in 1977. Both areas experienced a net loss in manufacturing employment during the 1967-79 period, despite the 8.5 percent increase nationwide. The areas with the highest rate of increase in employment, the Mountain and West South Central states, between them were responsible for about a ninth of the value added by manufacturing in 1977. Moreover, there has been a steady decline in the proportion of total employment in manufacturing in the Northeast and North Central regions, especially in durable goods manufacturing, and a substantial increase in the share of this employment in the South and West Table 5–13).

Important structural changes have been taking place in manufacturing in the recent past. Some industries, primarily those concerned with energy or high-technology products such as aerospace or microprocessors, have expanded rapidly in the past decade. The regions in which these industries are concentrated, mainly the sunbelt states, have become the high growth areas. The sectors that have been growing slowly or declining include many of the mature industries in which the United States was the world leader for many years. They include the basic steel industry, metal products (e.g., containers), appliances and trucks. The slowest growing regions in the past decade have been in the East North Central and Middle Atlantic states where these older industries have been concentrated.

Table 5–13. Distribution of Employment in Durable Goods Industries, by Major Region, 1968-78.

Region	Percentage of Durable Goods Industries Total Employment	
	1968	1978
Northeast	28.7	25.3
North Central	37.4	36.6
South	19.2	22.8
West	14.7	15.3
Total	100.0	100.0

Source: Monthly Labor Review (1980: 15)

Twelve states in which nearly 200,000 or more persons held jobs as production workers in metalworking (SIC 33-38) in 1977 are shown in Table 5–14.) More than half of all metalworking production workers were employed in the five Great Lakes states plus California and New York. The twelves states listed employed over 70 percent of all production workers in these industries.

More than in most of the other states listed, employment in metalworking industries in California is concentrated in aerospace and electronics, two of the industries in the metalworking sector with the best growth prospects. In the five Great Lakes states, Illinois, Indiana, Ohio, Michigan, and Wisconsin, at least 56 percent or more of the production workers in manufacturing were

Table 5–14. Employment of Production Workers in Metalworking: Leading States, 1977.

State	Division	Metalworking Production Workers, 1977 SIC 33-38 (in 000s)	Metalworking Production Workers as Percentage of All Manufacturing Production Workers
United States total		6,313.7	46.1 percent
Ohio	East North Central	601.2	65.0
Michigan	East North Central	591.1	74.9
California	Pacific	567.9	49.7
Illinois	East North Central	503.0	58.6
Pennsylvania	Middle Atlantic	457.6	49.0
New York	Middle Atlantic	390.3	40.7
Indiana	East North Central	347.8	66.6
7 state total		3212.2	50.8 of U.S. total
Texas	West South Central	258.5	43.0
Wisconsin	East North Central	211.7	55.6
Massachusetts	New England	196.0	48.0
New Jersey	Middle Atlantic	186.6	38.1
Missouri	West North Central	145.4	48.6
12 state total		4,457.1	71.0 of U.S. total

Source: Census of Manufacturers (1981b: Table 5).

employed in the metalworking industries. These states, and especially Michigan, are likely to experience greater than proportionate impacts of robotization on employment in the next decade.

It is likely that regional shifts in manufacturing employment will continue during the next decade. Lower labor costs, cheaper energy, lower taxes, and largely nonunion labor supply add up to what is widely regarded as a more favorable business climate in the sunbelt. This factor is frequently supplemented by lower transportation costs arising from proximity to markets and sources of raw materials. The Fantus Company, a location consulting firm for large corporations, recently assessed the business climate in the 48 contiguous states. Seven of the ten states regarded as possessing the most favorable business climate were in the South. Seven of the ten states rated as possessing the least favorable climate for business were in the Northeast or North Central regions.

It can be argued that tax policies intended to encourage business firms to build new plants may have the side effect of speeding up regional shifts in manufacturing activity. Changes in corporate taxes, such as more rapid depreciation allowances for plant and equipment, can increase productivity and the pace of GNP growth for the entire nation. These changes might favor the construction of new plants designed to take advantage of computer-assisted manufacturing, including robotization, in comparison to upgrading older plants. By reducing the after-tax cost of building a new plant elsewhere, they might hasten the process by which the North Central and Middle Atlantic states have been losing their manufacturing base. This is a question that deserves further study.

The regional patterns of economic and employment growth roughly coincide with similar patterns of regional migration. The Northeast and North Central regions have experienced a net outmigration while the South and the West have experienced net population increases (Table 5–15).

While some persons have moved for noneconomic reasons, such as the desire for a change of climate, the primary reason in a majority of instances has probably been related to employment. On an average, the regional migrants in the 1970s have been better educated and younger than the nonmigrants. The highest rate of interstate migration occurs at age twenty-three, an age that cor-

Table 5–15.   Persons Moving to and from Each Major Region, 1973-76.

| Major Region | Migrants (in 000s) | | Ratio of Outmigration to Inmigration |
	Into Region	Out of Region	
Northeast	1,058	1,829	173.0
North Central	1,935	2,400	124.0
South	3,254	2,907	74.0
West	2,100	1,718	81.5

Source: Bureau of the Census (1979).

responds closely to the usual age of graduation from college. The propensity of more highly educated persons to migrate to the South and West contributes to the development of high technologies in these regions. It helps to assure employers of an adequate supply of skilled engineers, scientists, systems specialists, technicians, and managers if they establish automated and robotized production facilities. The significance of the regional shifts is that they tend to magnify the impact of job losses due to robotization. The snowbelt regions that will lose most of the jobs taken over by robots are also losing jobs, in general, to the sunbelt.

UNION RESPONSE TO TECHNOLOGICAL CHANGE

The industries that are candidates for extensive robotization are mostly characterized by the presence of strong unions and well-established collective bargaining procedures. Over one-third of all wage and salary workers in manufacturing, and a significantly higher proportion of production workers—85 percent of motor vehicle equipment operatives and 41 percent of nondurable goods operatives—are represented by labor organizations. Over 90 percent of those represented actually belong to unions (Table 5–16). Policies for dealing with the displacement, therefore, are unlikely to be adopted unless organized labor participates in their formation. The major unions representing workers in the metalworking industries are listed in Table 5–17). There are no reliable statis-

Table 5–16. Wage and Salary Workers Represented by Labor Organizations, May 1980.

	Percentage of Employed Wage and Salary Workers Represented by Labor Organizations	Number of Employed Wage and Salary Workers Represented by Labor Organizations (000s)	Number of Represented Workers in Unions (000s)
All Occupations and Industries	25.7	22,493	20,095
MANUFACTURING OCCUPATIONS			
Machinist and job setters	56.9	397	381
Other metalworking craft workers	63.1	423	411
Motor vehicle equipment operatives	85.8	315	312
Other durable goods operatives	46.8	1,917	1,802
Nondurable goods operatives	40.8	1,320	1,244
Manufacturing laborers	52.2	436	420
MANUFACTURING INDUSTRIES			
Manufacturing, total	34.8	7,309	6,771
Durable goods, total	37.6	4,720	4,366
Ordinance	20.9	86	74
Lumber	20.9	113	103
Furniture	28.6	132	124
Stone, clay and glass	49.4	305	292
Primary metals	60.5	712	686
Fabricated metals	39.0	530	491
Machinery, exp. electrical	30.6	851	798
Electrical machinery	30.1	672	599
Transportation equipment	55.9	1,135	1,038
Automobiles	63.1	600	582
Aircraft	50.4	341	286
Other transportation equipment	48.1	194	170
Instruments	14.5	90	79
Miscellaneous	18.8	93	82
Nondurable goods	30.7	2,589	2,405

Sources: Bureau of Labor Stastics (1981a).

Table 5–17. Major Unions Representing Workers in the Metalworking Industries.

Union	Membership 1978 (000s)	1980 (000s)
United Automobile, Aerospace and Implement Workers of America (UAW)	1,499	1,357
United Steel Workers of America (USW)	1,286	1,238
International Brotherhood of Electrical Workers (IBEW)	1,012	1,041
International Association of Machinist and Aerospace Workers (IAM)	724	754
International Union of Electrical, Radio, and Machine Workers (IUE)	255	233
United Electrical, Radio and Machine Workers of America (UE)	166	162

Sources: 1978 membership: Bureau of labor Statistics (1979c), 1980 membership: Bureau of Labor Statistics (1980a)

tics that cross-classify union membership by manufacturing industry, but it appears that almost all of the membership of the United Auto Workers (UAW), International Association of Machinists (IAM), International Union of Electrical Workers (IUE), United Electrical Workers (UE), and United Steel Workers (USW) are in the metalworking industries (sectors 33-38). On the other hand, most of the membership of the International Brotherhood of Electrical Workers (IBEW) work outside of manufacturing.

Labor organizations in the past have shown only a moderate concern with robots. At first, robots were used mainly in dirty, monotonous, and unsafe jobs. The general assumption by unions in the past, and one supported by robot manufacturers in promoting their use, was that the employees who were displaced could readily be absorbed in job openings created by normal attrition or by growth. This assumption was reinforced somewhat by the loss of credibility of earlier forecasts of mass displacement of blue-collar workers due to automation. However, exceptionally rapid growth in the use of industrial robots since 1979, despite a general economic slowdown, has rekindled workers' fears about loss of jobs.

As of 1980, between two-thirds and three-fourths of operatives and laborers were less than forty-five years old, which means that barely a third of these workers would be retired in the normal way by the year 2000 (Table 5–18). On the average, skilled workers are older but they are not as likely to be replaced by robots in the near future. This suggests the possibility of retiring some workers earlier than normal. However, provisions for early retirement are less likely to figure as part of the solution to the robot displacement problem in the 1990s than in the earlier collective bargaining agreements dealing with technological change. Demographic changes have been increasing the number of older persons in the labor force, while early retirements have increased the proportion of retirees, in the unions that would mainly be affected. The United Automobile Workers Union, for example, included 190,000 retirees from the Big Three automobile manufacturers among its members in the late 1970s (Business Week 1979). Faced with escalating Social Security taxes and costs, national policy has been shifting from favoring early retirement as an employment-creating system for young persons toward favoring proposals to keep more older persons in the labor force. Recent proposals to raise the age for qualifying for full Social Security benefits from sixty-five to sixty-eight and the 1978 amendments to the Age Discrimination Act outlawing compulsory retirement for most retirees before age 70 symbolize the shift in public policy. Moreover, dependence on company pensions will become less attractive to employees if inflation continues since they are seldom indexed to changes in the cost of living. Emphasis on retirement gains also raises the possibility of creating intergenerational conflicts within the unions, since the gains for the employed members must be traded off for inflation adjustments or other benefits for older persons who are already retired or who are about to retire. Unions can be expected to favor two approaches in dealing with the job losses for their members threatened by robotization. One is to transfer and retrain the displaced employees into other jobs that have been created by attrition or by

Table 5–18. Age Distribution of the Manufacturing Workforce, 1980.

Occupation	Number Employed (000s)	Percentage Distribution by Age Group								Percentage 45 or younger
		16-19	20-24	25-34	35-44	45-54	55-59	60-64	65+	
Total Employed	97,290	7.8	14.0	27.0	19.8	16.7	7.2	4.5	3.0	69.0
Machine jobsetters	658	3.3	15.2	27.8	19.6	17.6	9.4	5.8	1.3	66.0
Other metalworking craft workers	638	2.0	9.6	28.8	20.8	20.0	11.0	6.7	1.1	61.0
Motor vehicle equipment operatives	431	2.1	11.3	30.6	25.9	20.1	5.8	3.9	.2	70.0
Other durable goods operatives	4,166	5.5	17.5	27.9	19.5	16.6	7.4	4.5	1.1	70.0
Nondurable goods operatives	3,290	6.5	16.0	26.3	19.3	18.5	7.5	4.5	1.4	68.0
Manufacturing laborers	961	9.7	20.3	28.2	16.5	14.1	5.9	3.8	1.4	75.0

Employment estimates are annual averages for 1980.
Source: Derived from unpublished data supplied by the Bureau of Labor Statistics, Employment Structure and Trends Division.

growth. This type of remedy is likely to be least costly to employers and to constitute a minimum barrier to the introduction of robots. The limitation of this approach is that it assumes a pace of robotization that is consistent with the number of suitable job openings created within the same plant (or in other plants belonging to the firm). This may not be possible in small firms or in declining industries.

The other approach favored by many unions is to attempt to protect threatened jobs by raising the cost of introducing new technology and in this way transferring part of the productivity gain from employers to employees. A side effect of this type of measure is to slow down the introduction of new technology, thus adversely affecting international competitiveness. Policies in this category include restrictive work rules, shortening the work week, lengthening paid vacations, or adding paid personal holidays. They also include employment guarantees and employer-financed pensions for older employees who retire early. Another measure intended to assure that workers' interests are taken into account in the decisions affecting job displacement is a requirement for advance notice to be given to unions before the new technology can be introduced.

Collective bargaining contracts are the major mechanism available to unions to affect conditions in the workplace. Past experience in collective bargaining offers an indication of the proposals that will be advanced by unions to deal with potential displacement caused by robotization. Typical collective bargaining provisions relevant to the introduction of new technology are shown in Table 5–19). Different types of provisions are briefly discussed, based on a review of typical recent (1980-81) contracts of the UAW, IAM, IUE, and IBEW.[9]

● Provisions calling for the *sharing of productivity benefits* are based on the assumption that technological improvements that increase productivity should in turn increase corporate profits. The UAW's wage improvement factor is an example of a clause calling for an annual percentage "productivity increase" exclusive of cost of living increases.

---

[9] For a more comprehensive overview of collective bargaining agreements, see AFL-CIO 1982).

Table 5–19. Collective Bargaining Provisions Relevant to Technological Change.

Type of Provisions	Specific Clauses
Advanced notice provisions	Layoffs Plant shutdown or relocation Technical change
Interplant transfer and relocation allowance provisions	Interplant transfer provisions Preferential hiring Relocation allowance
Unemployment compensation provisions	Supplemental unemployment benefit plans Severance pay Wage-employment guarantee
Seniority and recall related provisions	Retention of older workers Merging seniority lists Retention of seniority in lay-off
Exclusion from job security provisions	Exclusions from job security grievence procedure Exclusion from job security arbitration procedure
Work sharing provisions applicable in slack work periods	Division of Work Reduction of hours Regulation of overtime
Education and training provisions	Leaves of absence for education Apprenticeship On-the-job training Tuition aid for training
Provisions calling for joint labor-management committees	Industrial relations issues Productivity issues

Source: Bureau of Labor Statistics (1981b).

- Paid personal holidays (PPH) are intended to spread fewer available jobs among a greater number of employees by giving workers additional days off with pay in addition to holidays. The UAW contract called for twenty-six paid personal holidays over a three-year period for each member working for an automobile manufacturer (about 50 percent of the UAW membership). The intent was to reduce the number of workers laid off by reducing the number of days worked per employee. Other union contracts have moved in this direction by increasing standard vacation time. Paid personal holidays have been retracted or reduced in the recent agreements signed between the UAW and the major auto makers.

- Supplemental unemployment benefits are used in addition to unemployment compensation to aid workers through layoff periods. Nationally, the UAW is the principal advocate of this program.

- Transitional Allowances are provided to workers under some contracts when the firm transfers employees from plant to plant. These allowances ranged from $500 to $1,760 per employee in four contracts we reviewed. In some cases, benefits will also follow transferred employees. Seniority does not transfer for the IAM.

- Advance notice of technological change is often required. The extent of union input and involvement varies among the major unions. Some UAW and IAM contracts specify committees to study and discuss each change in technology with both union and management representation, whereas management retains the sole right of controlling the introduction of new technology in IBEW plants.

- Severance pay is a lump-sum compensation for workers who are permanently laid off. This is sometimes called "special retirement," where workers are paid lump sums to leave the job, in addition to receiving a percentage of their original pension benefits. This plan is a quick, but costly, means of reducing the size of the workforce.

- The integrity of the bargaining unit has been an issue in recent collective bargaining agreements. The UAW has several agreements stating that all jobs previously in the bargaining unit will stay in the unit. In other words, if an operative in a bargaining unit is replaced by a robot, then the robot's operator will also be in the unit.

Within the metalworking industries, three of the major unions, the UAW, the IAM, and the IUE, have agreements with their employers providing for joint-management involvement in committees concerned with the impact of technological change for employees. Collective bargaining agreements for at least three of these unions, the UAW, IAM, and IUE, also include provisions for retraining displaced workers, with the costs borne by the employer.

Many of the measures favored by unions in the past to deal with the loss of jobs due to technological advances are unlikely to be available in the coming decade to the unions in the metalworking industries. A frequent union response in the past has been to trade the displacement of some employees in exchange for work guarantees for employees who were retained. The collective bargaining agreements between the longshoremen and the shipping industry at the time containerization was first introduced in the 1950s has set a precedent for this type of arrangement. The employers, in effect, "buy out" the opposition to the new technology by sharing the productivity gains with the employees who are affected through work guarantees, pensions, and increases in wage rates. The coal miners' union permitted mechanization of mines in exchange for the creation of a pension fund paid for by a fixed charge on each ton of coal mined. Similarly, the legislation establishing CONRAIL in the early 1970s provided that employees displaced because of the elimination of railway passenger runs would receive allowances, varying with their length of service that could continue up to age sixty-five. The Typographical Union has responded to the computerization of typesetting by requiring that its members be retained to work the new machines. In many areas New York, for example—this demand has been coupled with employment guarantees assuring that the reductions in force made possible by computerization be achieved through attrition.

While unions such as the UAW often mention employment guarantees in their demands, the arrangements involving employment guarantees are less applicable to semiskilled workers in mass production industries such as motor vehicles or steel than they are in some other industries; for example, railroads. The labor market in which these unions operate is currently unfavorable to labor protection involving substantial increases in costs. Employers in the automobile or electronic equipment industry have been responding to increases in labor costs by "outsourcing" more parts from nonunion or foreign suppliers. Indeed, in recent years, the loss of union jobs has virtually become a hemorrhage. These prospects set limits to the bargaining power of unions to cope with robotization and related advances in the 1980s.

JOB OPENINGS

In the occupations expected to be primarily affected by robotization, the job openings likely to be created by attrition in the 1980s provide a basis for assessing the policies for dealing with displacement in the next decade or two. Attrition rates in the operatives occupations are approximately 3 percent a year or slightly less, depending on the sex and age distribution of the persons employed in them. However, these figures substantially underestimate the number of people transferring out of specific occupations, since they only include people who leave the establishment.[10] A 3 percent per annum attrition rate suggests an annual average of about 170,000 job openings among blue-collar workers in the metalworking industries alone during the 1980s.[11] If one-half of these openings could be filled by operatives displaced from other jobs by robots, an average of over 85,000 a year, or nearly 850,000 positions could be filled during the decade, while still accommodating some other job seekers. The rate of movement out of specific occupations appears to be much higher than is indicated by average industry wide attrition rates which only measure movement out of the establishment, as shown in Table 5–20. These estimates of movements out of occupations are based on a recently compiled Bureau of Labor Statistics (1980b) study using data collected between 1973-76.

Only 82 percent of the people employed as welders in the first year of the survey were still employed as welders in the next year. Of the 18 percent who moved out of welding jobs during the next year, nearly 8 percent were still employed but in different occupations. The remaining 10 percent were either unemployed or had left the labor force (deaths, retirements, disability, etc.). These figures suggest that there is a high rate of annual labor turnover (greater than 20 percent) within several of the semiskilled and unskilled types of jobs that could be robotized.

---

[10] Attrition is used to refer to workers who leave the establishment as a result of quits, discharges, permanent disability, death, retirement, or transfers to other companies. The other main source of labor turnover is layoffs (suspensions without pay for more than seven consecutive days initiated by the employer). Together, the attrition rate and the layoff rate comprise the "total separation" rate.

[11] Assuming 6.7 million blue collar workers in metalworking (Table 5–11), and an average rate of 3 percent decrease per year over a ten-year period.

Table 5–20.   Movements Out of Occupational Groups.

	BASE YEAR	NEXT YEAR	
	Percentage Employed in Specific Occupation	Percentage Employed in Same Occupation	Percentage Employed in Different Occupation
Major Occupation Groups			
Craft and kindred workers	100.0	85.4	6.0
Operatives,except transport	100.0	73.4	11.4
Nonfarm laborers	100.0	81.4	14.5
Selected Occupations			
Machinist	100.0	94.1	4.1
Assemblers	100.0	72.6	13.2
Packers, wrappers, except meat and produce	100.0	67.9	15.5
Welders and flame cutters	100.0	82.1	7.6
Freight, material handlers	100.0	70.1	13.1

Source: Bureau of Labor Stastics (1980b: Table 1).

As the number of new entrants into the labor force declines in the 1980s because of the drop in birth rates after the mid-1960s, older and more experienced blue-collar job seekers will face less competition from younger workers for these openings than was the case in the 1970s. However, an unresolved question, at this point, is the extent to which economic growth or continued recession in basic industries such as automobile and steel will increase the numbers of job seekers competing with employees displaced because of robotization. In a declining industry, moreover, openings created by attrition are often left unfilled. Turnover openings that would arise from occupational mobility can fall off sharply as few persons are added to the employment roll, and few quit voluntarily to take other positions.

In an economy in which many industries are growing while others are declining, there is a choice of occupations suitable for retraining programs. The Bureau of Labor Statistics has identified a number of industries in which high rates of job growth and/or large numbers of job openings are anticipated in the coming decade and probably beyond. The list (Table 5–21) illustrates the fields in which the persons displaced from operative jobs could hope to find employment with or without additional training, though not necessarily at equal or higher pay. The BLS projections assume that the drift toward a post-

Table 5–21. Projected Employment Changes for High Growth Industries, 1979-90.

Fastest Growing Industries	Average Annual Rate of Job Growth (Percentage)
Other medical services	4.6
Typewriters and other office equipment	4.5
Computers and peripheral equipment	4.2
Coal mining	4.1
Hospitals	3.8
Crude petroleum and natural gas	3.6
Doctors' and dentists' services	3.4
Local government passenger transit	3.3
Other state and local government enterprises	3.2
Automobile repair	3.1

Industries with Largest Job Gains	Employment Gain (000s)
Eating and drinking places	1,912
Retail trade, except eating and drinking places	1,878
Hospitals	1,347
Miscellaneous business services	1,171
Other medical services	909
New construction	892
Wholesale trade	866
Doctors' and dentists' services	580
Banking	490
Educational services (private)	416

Projections based on the Bureau of Labor Statistics low-growth-trend forecast. Source: Personik (1980: 40).

industrial society will continue. Most of the sectors projected to have large numbers of employment openings are in areas other than manufacturing, though two manufacturers in the metal-working sector are among those projected to experience an above average employment growth. A common feature of most, but not all, of the most rapidly growing occupations is that they presuppose specialized training or education beyond the high school level.

Many of the blue-collar workers displaced by robots do possess the educational qualifications for more skill training, which could lead to a better paid position in other occupations. The tradition-al stereotype of a factory operative has been that of a person with limited education, often a "functional illiterate." A genera-tion ago, the typical educational attainment level for operatives was below that of the overall workforce. In the mid-1950s, for in-stance, the median number of years of schooling completed by operatives was 9.5 years as compared with 11.7 years for all em-ployed civilian workers. By 1978, the median for operatives, ex-cluding transport operatives, was 12.1 years. This compared with an overall average of 12.6 years. Operatives, as a group, tend to possess a high school education or its equivalent. This can pro-vide a basis for further specific vocational training or for further higher education in a two-year or four-year college.

Over time, a shift toward flexible automation of production could itself be an important factor in increasing the numbers and kinds of jobs that could provide employment for displaced and semiskilled workers from metalworking industries. Some of the employees displaced by the new technology may find work in the capital goods industries producing robots and related equipment. Indeed, the state of Michigan has launched a formal program to attract robot manufacturers to the state University of Michigan 1982). The greater productivity brought about by more widespread use of robots could also increase the competi-tiveness of some U.S. manufacturing industries in domestic and world markets sufficiently to reabsorb some of the employees who had lost their jobs to robots. Many new jobs would be cre-ated in consumer goods industries as greater productivity growth in the economy led to increases in personal income and consumer spending. Much of the indirect effects of robotization in increasing employment would be felt in service industries, including the business services that maintain and repair robots and similar equipment.

Changes in national priorities could also expand the range of job openings and outmode projections of the Bureau of Labor Statistics type, based on the experience of the recent past. A shift in national priorities favoring more adequate home care, income support, and medical services for the elderly, the retarded, and

Table 5–22. Projected Employment Changes for High Growth Occupations, 1978-90.

Most Rapidly Growing Occupations	Percentage Growth in Employment, 1978-90	Occupations with Largest Increases	Growth in Employment (000s)
Data processing machine mechanics	147.6	Janitors and sextons	671.2
Paralegal personnel	132.4	Nurses' aides and orderlies	594.0
Computer systems analysts	107.8	Sales clerks	590.7
Computer operators	87.9	Cashiers	545.5
Office machine and cash register servicers	80.8	Waiters/waitresses	531.9
Aero-astronautic engineers	73.6	General clerks, office	529.8
Food preparation and service workers, fast food restaurants	68.8	Professional nurses	515.8
Employment interviewers	66.6	Food preparation and service workers, fast food restaurants	491.9
Tax preparers	64.5	Secretaries	487.8
Correction officials and jailers	60.3	Truck drivers	437.6
Architects	60.2	Kitchen helpers	300.6
Dental hygienists	57.9	Elementary schoolteachers	272.8
Physical therapists	57.6	Typists	262.1
Dental assistants	57.5	Accountants and auditors	254.2
Peripheral EDP equipment operators	57.3	Helpers, trades	232.5
Child-care attendants	56.3	Blue-collar supervisors	221.1
Veterinarians	56.1	Licensed practical nurses	215.6
Travel agents and accommodations appraisers	55.6	Guards and doorkeepers	209.9
Nurses' aides and orderlies	54.6	Automotive mechanics	205.3

Projections based on the Bureau of Labor Statistics low-growth-trend forecast.
Source: Cary (1981b: 48).

the handicapped would be reflected in new kinds of jobs for persons with the appropriate retraining. Private and public efforts to rehabilitate physical infrastructure (bridges, subways, water/sewer systems, etc.)—especially in the Northeast and Great Lake states could create large numbers of job openings that could partially be filled by unemployed automobile workers from Detroit. To illustrate the possibilities, engineering standards recommend a replacement cycle for city streets in large cities approximating forty years. Continuing the recent actual replacement rates in New York City would imply a 150- or 200-year replacement cycle (Committee on the Future 1978). Clearly, a backlog of potential future demand for construction work is building up.

Looking ahead to the 1990s and beyond, two major developments are likely to influence substantially the distribution of the productivity gains from robotized and related computer-assisted technology. One is the changing demographic profile as the baby-boom generation of the 1950s and 1960s reaches retirement age in the first quarter of the next century. The other is the likelihood of gradual but significant reductions in annual work hours and the standard work week in manufacturing industries over the next two or three decades.

By the year 2010 there will be an estimated 10.8 million more persons sixty-five and over in the population than in 1978. With present retirement practices continuing for another two or three decades, the ratio of the employed workforce to the number of retired persons would decrease from approximately 3 to 1 in the late 1970s to 2 to 1 in the next generation. One recent report estimates the cumulative costs of the current types of federally funded programs that benefit persons sixty-five and over at $5.9 trillion by the year 2025 (Storey and Hendricks 1979). Allowing for the likelihood that in the future a longer lived and healthier population will work more years than in the recent past, it is probable that Social Security and other pension system costs will absorb much of the growth in manufacturing productivity attributable to new technology.

The forty-hour standard work week has remained unchanged in most manufacturing industries for the past twenty years. White-collar workers typically enjoy a shorter work week. For example, two-thirds of all office workers in the finance, insurance, and real estate sectors now work a standard work week of less than 40 hours. A gradual reduction in the standard work week would diminish job losses by spreading the available work over more employees and levying penalty rates for overtime after fewer work hours. However, the reduction in annual work hours could be taken in various ways. Sabbaticals, now confined to teachers and to some civil servants and steel workers, could be extended generally to production workers. The sabbaticals could be used to explore another occupation, to travel, to rest at home, or to become a student again. Blue-collar workers returning to or first entering a university because they were on a sabbatical could provide a new market for the services of colleges and universities faced with shrinking enrollments because of low birth rates two decades earlier.

Technological changes such as robotization are essential ingredients in the gale of creative destruction in which new industries come into being and old ones become rejuvenated or decline. While these changes could create serious job losses for semiskilled factory workers, mainly in metalworking industries, it need not follow that their overall effect would be to reduce employment. However, there is a strong likelihood of displacement and unemployment among certain categories of industrial workers, especially in the older manufacturing centers.

Many of the workers who lose their jobs can be expected to resolve their personal employment problem independently of the agreements reached between unions and management by relocating at their own expense, often by making use of severance pay or unemployment benefits for this purpose. The recent newspaper reports of the movement of auto workers from Detroit to places such as Houston suggest that this solution is still relevant. The problem with this remedy is that the costs of technological change are very largely and unfairly borne by the displaced workers themselves. These costs include relocation and job search costs, losses in the sale of homes in a depressed community, and the likelihood that the new job will pay less to semiskilled workers in a new and generally nonunion environment in which they lack seniority. If the numbers of displacees are not too great, our diverse economy will eventually absorb them. However, continued economic weakness could create a new class of dispossessed immigrant industrial "Okies," carrying with them a large potential for social unrest and political backlash against the "system" that failed to protect them. We return to the policy implications in Chapter 7.

Table 5–23.   Definitions of Skill Levels for a Job's Relationship to *Things*

Things: Inanimate objects as distinguished from human beings, substances or materials; machines, tools, equipment, and products. A thing is tangible and has shape, form, and other physical characteristics.

0 *Setting up*: Adjusting machines or equipment by replacing or altering tools, jigs, fixtures, and attachments to prepare them to perform their functions, change their performance, or restore their proper functioning if they break down. Workers who set up one or a number of machines for other workers or who set up and personally operate a variety of machines are included here.

1. *Precision working*: Using body members and/or tools or work aids to work, move, guide, or place objects or materials in situations where ultimate responsibility for the attainment of standards occurs and selection of appropriate tools, objects, or materials, and the adjustment of the tool to the task require exercise of considerable judgment.

2. *Operating-controlling*:Starting, stopping, controlling, and adjusting the progress of machines or equipment. Operating machines involves setting up and adjusting the machine or material(s) as the work progresses. Controlling involves observing gauges, dials, and so forth, and turning valves and other devices to regulate factors such as temperature, pressure, flow of liquids, speed of pumps, and reactions of materials.

3. *Driving-operating*: Starting, stopping, and controlling the actions of machines or equipment for which a course must be steered, or which must be guided, in order to fabricate, process, and/or move things or people. Involves such activities as observing gauges and dials; estimating distances and determining speed and direction of other objects; turning cranks and wheels; pushing or pulling gear lifts or levers. Includes such machines as cranes, conveyor systems, tractors, furnace charging machines, paving machines and hoisting machines. Excludes manually powered machines, such as handtrucks and dollies, and power assisted machines, such as electric wheelbarrows and handtrucks.

Table 5–23.   Continued

---

4. *Manipulating*: Using body members, tools, or special devices to work, move, guide, or place objects or materials. Involves some latitude for judgment with regard to precision attained and selecting appropriate tool, object, or material, although this is readily manifest.

5. *Tending*: Starting, stopping, and observing the functioning of machines and equipment. Involves adjusting materials or controls of the machine, such as changing guides, adjusting timers and temperature gauges, turning valves to allow flow of materials, and flipping switches in response to lights. Little judgment is involved in making these adjustments.

6. *Feeding-offbearing*: Inserting, throwing, dumping, or placing materials in or removing them from machines or equipment which are automatic or tended or operated by other workers.

7. *Handling*: Using body members, handtools, and/or special devices to work, move or carry objects or materials. Involves little or no latitude for judgment with regard to attainment of standards or in selecting appropriate tool, object, or material.

Source: Employment and Training Administration (1977: 1370–71).

Table 5–24. Examples of Range of Physical Skill levels for Operatives.

ELECTRICAL AND ELECTRONICS EQUIPMENT (SIC 36)

SKILL LEVELS

Occupations	0	1	2	3	4	5	6	7
Assemblers		Transformer assembler	Governor assembler		Galvanometer assembler	Light bulb assembler		Variable capacitor assembler
Operators	Machine operator, all around	Rework operator	Wafer machine operators		Diffusion furnace operator	Etch tank operators	Twisting machines operator	Seal mixing operator
Chippers Cleaners, and/or polishers						Bottom polisher		Tank cleaner
Inspector		Transformer Inspector			Resistor inspector	Weld inspector		Casting inspector
Helpers' Trades Painters, Production						Impregnant-ing helper		Lead burner helper Painter battery brand vent plug
Testers		Electronics Test I			Battery tester			Polarity tester

TRANSPORTATION EQUIPMENT (SIC 37)

Assemblers		Custom frame assembler		Assembler, axle		Assembler, oil filter
Operators	Drivematic machine operator	Roadability machine operator	Wheelpress operator	Steamtank operator	Core-Saw operator	Dielectric Press operator
Chippers Cleaners, and/or polishers				Car cleaner		Cardapron cleaner
Inspector		Motor & chassis inspector		Carburetor Inspector		Inspectors, oil filter
Helpers, trades						Motor test helper
Painters, production		Painter automotive		Painter insignia		Painter chassis
Testers		Dynamometer Tester	Hypoidgear Tester	Transmission tester		

Source: Derived from information given in Employment and Training Administration (1977).

Table 5–25.   Supplementary Skill Classifications for the Dictionary of Occupational Titles.

Physical demands	Strength (sedentary, light, medium, heavy, very heavy)
	Climbing and/or balancing
	Stepping, kneeling, crouching, and/or crawling
	Reaching, handling, fingering and/or feeding
	Talking and/or feeling
	Seeing (acuity, depth perception, field of vision, accommodation)
Environmental conditions	Inside, outside or both
	Extreme of cold plus temperature changes
	Extreme of heat plus temperature changes
	Wet and humid
	Noise and vibrations
	Hazards, fumes, odor, toxic conditions, dust, and poor ventilation
Mathematical development	Advanced calculus
	Algebra
	Computing discount, interest, profit, loss
	Add, subtract, multiply and divide all units of measure
	Add, subtract two-digit numbers
Language development	Types of reading required
	Types of writing required
	Types of speaking required
Specific vocational preparation	*Type of training*
	Vocation training
	Apprentice training
	In-plant training
	On-the-job training
	Eessential experience in other jobs
	Length of training
	Short demonstration–1 month
	1-3 months
	3-6 months
	6 months-1 year
	1-2 years
	2-4 years
	4-10 years

Source: Employment and Training Administration (1981: 465–74).

238   ROBOTICS: APPLICATIONS AND SOCIAL IMPLICATIONS

REFERENCES

Industrial Union Department, AFL-CIO. *Comparative Survey of Major Collective Bargaining Agreements: Manufacturing and Non-Manufacturing.* Technical Report, AFL-CIO, January 1982.

Ayres, Robert U., and Steven M. Miller. Robotics, CAM and Industrial Productivity. *National Productivity Review* 1(1), Winter 1982.

Ayres, Robert U., and Steven M. Miller. Industrial Robots on the Line. *Technology Review* 85(4):35–45, May/June 1982.

Bureau of Labor Statistics, U.S. Department of Labor. *Employment Projections for the 1980's.* Government Printing Office, Washington, D.C., 1979. Bulletin 2030.

Bureau of Labor Statistics, U.S. Department of Labor. *Metal Working Occupations.* Government Printing Office, Washington, D.C., 1979. Bulletin No. 2075–2, Reprinted from the Occupational Outlook Handbook, 1980–81 edition.

Bureau of Labor Statistics, U.S. Department of Labor. *Directory of National Unions and Employee Associations.* Government Printing Office, Washington, D.C., 1979. Bulletin 2079.

Bureau of Labor Statistics, U.S. Department of Labor. Principal U.S. Labor Organizations. 1980. Unpublished data.

Bureau of Labor Statistics, U.S. Department of Labor. *Measuring Labor Force Movements: A New Approach.* Government Printing Office, Washington, D.C., 1980. Report 581.

Bureau of Labor Statistics, U.S. Department of Labor. *Earnings and Other Characteristics of Organized Workers, May 1980.* Government Printing Office, Washington, D.C., 1981. Bulletin 2105.

Bureau of Labor Statistics, U.S. Department of Labor. *Characteristics of Major Collective Bargaining Agreements: 1 January 1980.* Government Printing Office, Washington, D.C., 1981. Bulletin 2095.

Bureau of Labor Statistics, U.S. Department of Labor. *Technology and Labor in Four Industries: Meat Products, Foundaries, Metalworking Machinery and Electrical and Electronic Equipment.* Government Printing Office, Washington, D.C., 1982. Bulletin 2104.

Bureau of Labor Statistics, U.S. Department of Labor. *The National OES Based Industry-Occupation Matrix for 1980.* Government Printing Office, Washington, D.C., 1982.

Bureau of the Census, U.S. Department of Commerce. *Reasons for Interstate Migration.* Special Studies, Number 81, Series p–23, 1979.

Bureau of the Census, U.S. Department of Commerce. *Statistical Abstract of the United States, 1980.* Government Printing Office, Washington, D.C., 1980. 101th ed.

Bureau of the Census, Industry Division, U.S. Department of Commerce. *1977 Census of Manufacturers*. Volume MC77–SR–1: *General Summary*. Government Printing Office, 1981.

Bureau of the Census, Industry Division, U.S. Department of Commerce. *1977 Census of Manufacturers*. Volume MC77–A–Part 1 and 2: *Geographic Area Statistics*. Government Printing Office, 1981.

Business Week, 23 July, 1979.

Carnegie-Mellon University. *The Impacts of Robotics on the Workforce and Workplace*. Department of Engineering and Public Policy, Carnegie-Mellon University, Pittsburgh, Pa., 1981. A student project cosponsored by the Department of Engineering and Public Policy, the School of Urban and Public Affairs, and the College of Humanities and Social Sciences.

Cary, Max L. Three Paths to the Future: Occupational Projections, 1980-90. *Occupational Outlook Quarterly* 25(4):2–11, Winter 1981.

Cary, Max L. Occupational Employment Growth Through 1990. *Monthly Labor Review* 104(8):42–55, August 1981.

Committee on the Future. *Regional and Economic Strategies for the 1980's*. Technical Report, The Port Authority of New York and New Jersey, 1978. p. 28.

Conigliaro, Laura. *Robotics Presentation Institutional Investor's Conference: 28 May 1981*. Technical Report, Bache Halsey Stuart Shields, Inc., 19 June 1981.

Employment and Training Administration, U.S. Department of Labor. *Dictionary of Occupational Titles 4th ed.* Government Printing Office, 1977.

Employment and Training Administration, U.S. Department of Labor. *Selected Characteristics of Occupations Defined in the Dictionary of Occupational Titles*. Government Printing Office, 1981.

Employment and Training Administration, U.S. Department of Labor. *Employment and Training Report of the President*. Government Printing Office, 1981.

Engelberger, Joseph F. *Robotics in Practice*. American Management Association, New York, 1980.

Kalmbach, P., R. Kasiske, F. Manske, O. Micker, W. Pelull, and W. Wobbe-Ohlenburg. Robots' Effect on Production, Work and Employment. *Industrial Robot* 9(1):43–46, March 1982.

Lecht, Leonard. *Occupational Projections for National, State, and Local Areas*. Technical Report, National Commission on Employment and Unemployment Statistics, 1979.

Lublin, Joanne. Steel Collar Jobs: As the Robot Age Arrives Labor Seeks Protection Against Loss of Work. *Wall Street Journal:* Section 1, page 1, 26 October 1981.

Miller, Tim. The Coming Job Crunch. *National Journal* 14(20):865–869, May 15, 1982.

*Monthly Labor Review:*15, March 1980.

Where the Jobs Are—And Aren't. *Newsweek* XCVIII(21):88–92, November 1981.

Nof, Shimon Y., and Hannan Lechtman. Robot Time and Motion System Provides Means of Evaluating Alternate Robot Work Methods. *Industrial Engineering* 14(4):38–48, April 1982.

Organisation for Economic Cooperation and Development. *The Requirements of Automated Jobs.* OECD, Paris, 1965.

Personick, Valeria A. The Outlook for Industry Output and Employment Through 1990. *Monthly Labor Review* 104(8):28–41, August 1981.

Rothwell, Roy, and Walter Zegveld. *Technical Change and Employment.* St. Martin's Press, New York, 1979.

Saunders, Norman C. The U.S. Economy through 1990– An Update. *Monthly Labor Review* 104(8):18–27, August 1981.

Storey, James R., and Garry Hendricks. *Retirement Income Issues in an Aging Society.* The Urban Institute, 1979.

Memo. Technical Report, Research Department, United Auto Workers Union, 20 October 1980. Mimeo.

University of Michigan. *Automatic Factory Opportunities in Michigan: Robotic Systems.* Industrial Development Division, Institute of Science and Technology, The University of Michigan, Ann Arbor, Michigan, 1982.

Yonemoto, Kanji. The Socio-Economic Impacts of Industrial Robots in Japan. *Industrial Robot* 8(4):238–41, December 1981.

6 PRODUCTIVITY IMPACTS

Robotics and computer-aided manufacturing (CAM) technologies are sometimes cited as the key to reversing the decline in U.S. productivity growth, restoring the international competitiveness of American goods, and stimulating overall levels of economic growth. These are questions that deserve serious attention. This chapter introduces a framework for understanding and analyzing the issues. It is beyond our scope, and frankly beyond the current state of knowledge, to confirm or refute these claims unambiguously. However, we can explain why there is a large potential for improving productivity within those industries that are among the first wave of robot users, and we can roughly estimate some of the cost savings that might be realized.

We begin by reviewing recent productivity trends and current thinking on the relationships between capital and productivity. We discuss the comparative economics of small-, medium-, and large-volume production and point out why and where there are substantial potentials for productivity improvement within each of these "domains." We estimate how much cost saving might be realized if robotics and CAM technologies could make small- and medium-volume production closer in efficiency to mass production operations. Subsequently, we analyze the magnitude of cost saving that might be realized by substituting robots for workers and the potential for increasing output by increasing the utilization of machine tools. Our calculations suggest that the savings theoretically obtainable from increasing the utilization of capital equipment are at least an order of magnitude larger than savings that could be realized by eliminating a fraction of labor cost.

We identify several consequences of achieving dramatic increases in output per unit of capital and labor (or equivalently, of reducing unit cost) in robotic plants. For one thing, fewer plants will be needed to meet current levels of demand. If demand remains at current levels, additional facilities will not be needed, and the older, less productive facilities will most likely be shut down. To reap fully the benefits of robotics, managers will have to find ways of stimulating demand and utilizing the additional capacity. This requires making greater use of the flexibility of robotic production systems either to manufacture new, rapidly evolving products or to produce a wider range of products. Thus, we suggest that the full benefits of robotics will be realized only if there is a parallel emphasis on product innovation. Finally, we briefly discuss some of the broader economic impacts of improving productivity in the metalworking sector. It appears that the application of robotics to batch manufacturing could result in significantly reduced real cost for capital goods in relation to other factors of production. This could have ripple effects on the prices of manufactured goods throughout the economy and beneficial long-term effects on the rate of inflation and on the competitiveness of U.S. industry.

OVERVIEW OF RECENT PRODUCTIVITY TRENDS

There are four important points to be made here regarding productivity trends:

1. All recent major studies of productivity growth confirm that there has been a substantial decline in the aggregate growth rate of productivity since 1973, although evaluations of the extent of this decline and the relative importance of the factors contributing to it vary from study to study.

2. All major studies of productivity document a slowdown in the formation of capital and suggest that increased levels of technical activity and capital formation are an important part of reviving productivity growth.

3. Only part of the decline in the overall productivity growth rate can be directly attributed to the slowdown in capital formation, both in manufacturing and throughout other sectors of the economy. Many of the reasons for declining productivity may be beyond the realm of improvements in manufacturing technology.

4. Recommendations that more efficient capital is the key to resuming our current rates of economic growth must be interpreted in a new light because of the simultaneous but mutually interfering requirements of reducing cost and of increasing product variety and quality. The concept of efficiency must be broadened to include technological systems that can economically produce a diverse and changing mix of goods.

Documenting the Decline in Productivity Growth

Productivity is the relationship between output and one or more of the associated inputs used in the production process. Output can be related to a single input, such as labor, to form a partial productivity measure, or to all related inputs to form a measure of total factor productivity (TFP). Relating output to all associated inputs, as in the TFP measure, estimates the net savings in real costs per unit of input, or increases in productive efficiency. Two different estimates of the annual average percentage rate of growth of total factor productivity for the private business economy are shown in Table 6–1. Differences in the concepts and procedures used to measure and aggregate inputs and outputs account for the large differences between the estimates.[1]

---

[1] These differences often stir heated debates among economist on the "appropriate" way to measure productivity and on explaining the underlying sources of economic growth or decline. The measurement and explanation of productivity change is a far more complex problem then most people realize, and any figures cited should be viewed with caution. See

Table 6–1.   Estimates of Total Factor Productivity Growth Rates.

Period	Average Annual Percentage Rates of Growth (Private Business Economy)	
	Kendrick and Grossman	Fraumeni and Jorgenson
1948-53	3.4	1.66
1953-57	2.1	1.46
1957-60	2.1	1.13
1960-66	3.4	2.11
1966-69	1.5	0.04
1969-73	1.8	0.95
1973-76	0.7	−0.70
1973-79[a]	0.4	
1979-80[a]	−1.4	

[a] American Productivity Center estimates, based on the procedures developed by Kendrick and Grossman. See Sadler (1981: 6).

Sources: Kendrick and Grossman (1980: 29-35); Fraumeni and Jorgenson (1980: 171–76).

Kendrick calculates the ratios of real product to real factor costs in successive time periods and then converts the ratios to index numbers. He does not adjust capital and labor input measures for changes (mostly improvements) in quality, as do some other productivity researchers, which leaves a large portion of the rate of change of output unexplained by the growth rate of inputs. Consequently, his estimates of productivity changes tend to be higher than most others.

Another approach to measuring productivity, used by Jorgenson and coworkers, is to construct separate measures of the rate of growth of aggregate output and aggregate inputs. The difference between the rate of growth of output and input indices is defined as the rate of growth of "technical change." It measures that part of the growth of output that cannot be accounted for

NRC (1979) for a comprehensive overview of the concept of productivity. See *Survey of Current Business*, May 1972, for reprints of the "classical" debates between Denison and Grilliches and Jorgenson on the concepts and measurement of productivity. See Usher 1980 for an overview of the problems of measuring economic growth.

by the growth of capital, labor, or intermediate inputs, that is, the rate of growth of output, holding all inputs fixed. Recent estimates compiled by Fraumeni and Jorgenson are shown in the last column of Table 6–1. Since they adjust both capital and labor inputs for changes in quality, and use a different framework for combining inputs, their measures of productivity are significantly lower in absolute value than those of Kendrick and Grossman. However, both approaches show a sharp decline in productivity growth since 1973.

While the aggregate rate of productivity growth has sharply declined since 1973, a more detailed look across the separate sectors of the U.S. economy reveals interesting intersectoral differences (Figure 6–1). According to American Productivity Center estimates, despite the overall slowdown in the 1973-79 period, two sectors, communications and services, actually realized higher productivity growth rates than they did in the earlier 1965-73 period. Between these two periods, the productivity slowdown in the private goods industries was more severe than in the private service sector. During the 1979-80 period, productivity actually declined for the private business sector as a whole, especially in the goods producing industries, with the exception of mining. The rates also declined in the private service sector overall, despite improvements shown by transportation and utilities, because of the large decreases in trade and communications.

Productivity growth rates are declining throughout most sectors of the private business economy. It would be naive to expect new production technologies, such as robotics and CAM, to alleviate all our economic problems.

LABOR PRODUCTIVITY AND UNIT LABOR COST

High labor costs have come to characterize most advanced capitalist economies, and they have encouraged a shift of manufacturing, especially of standardized products, to less developed nations with low-cost labor. Multinational corporations have facilitated the development of manufacturing "export platforms" in Hong Kong, Singapore, Malaysia, Mexico, South Korea, or Taiwan, especially for electronics and apparel. The "world cars" will contain

Figure 6–1. Total Factor Productivity in the U.S. Economy, by Sector.

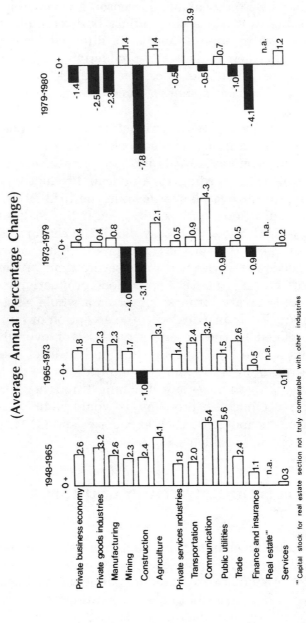

(Average Annual Percentage Change)

Source: Sadler (1981: 6).

parts drawn from subsidiary plants of American firms in such nations as Brazil, Mexico, or Spain—or imported from Japanese suppliers. Looking ahead, according to Peter Drucker (1980), "a labor
intensive manufacturing plant in an advanced industrial economy
is doomed to become noncompetitive fairly fast." We argue, however, that standardization of products in technology is the key to
emigration of manufacturing away from the United States.

Unit labor costs increase when hourly compensation outpaces
the growth of output per hour. Unit labor costs in U.S. manufacturing industries have increased by over 90 percent in the 1967-79
period. Money wages and fringe benefits increased by over 150
percent, far outpacing the 30 percent increase in output per worker (Table 6–2). Moreover, labor costs have been rising in U.S.
manufacturing industries at the same time that many of these
industries, such as steel, automobiles, and electronic and electrical
appliances, have been subjected to severe competition in their domestic markets from foreign firms (Table 6–3).

Employee compensation has also been rising in other major industrial nations shown in Table 6–4, including Japan and West

Table 6–2.   Output, Compensation, and Unit Labor Costs in
Manufacturing, 1950-79.

Year	Output per Employee Hour	(1967 = 100) Indices of Compensation Per Employee Hour[a]	Unit Labor Costs
1950	65.7	45.1	69.4
1960	79.4	77.1	97.7
1965	98.5	91.0	92.6
1967	100.0	100.0	100.0
1970	104.5	121.8	116.5
1975	118.2	180.2	152.4
1978	128.1	229.5	179.4
1979	130.3	250.5	192.4

[a] Wages and salaries of employees plus employer's contribution for social insurance
and private benefit plans.
Source: Bureau of Labor Statistics (1980b: 206).

Table 6–3.   Import Measures for Selected Metalworking Products.

Year	Black and White Televisions (A)	Color Televisions (A)	Semiconductors (B)	Autos (A)	Trucks (A)
1968	36.8	11.0	6.0	10.5	*
1970	79.1	18.9	11.0	14.7	*
1972	62.0	15.7	19.0	14.5	*
1974	78.4	16.3	29.0	15.7	8.5
1975	59.8	15.7	29.0	18.2	6.2
1976	83.2	36.8	28.0	14.8	8.2
1977	86.6	27.8	32.0	18.3	6.3
1978	97.8	27.1	31.0	17.8	8.0
1979	93.9	13.9	32.0	22.6	11.4
1980	a	a	a	26.7	19.5

Import Measure: (A) Imports as percent of total sales; (B) Imports as percent of production.
a Data not given
Source: Office of Technology Assessment (1981: 50-53).

Germany throughout the 1970s, but unit cost increases have been moderated by more rapid productivity growth in these countries over the 1973-80 period. Absolute levels of labor productivity, measured by gross domestic product per capita, are still higher in the United States than in most of its major trading partners, but the more rapid increase in productivity in other nations imply that the differential favoring American manufacturing industries continues to narrow. The average annual growth rate of output per hour in manufacturing industries in the United States in the 1973 to 1980 period was about a fourth of the comparable growth in Japan and Belgium, less than one-third of the rate in the Netherlands, and slightly more than a third of the increase in West Germany and France.

The Japanese have so far been the world's leaders in restraining rising labor costs and obtaining productivity increases. On a national currency basis, Japan has experienced the slowest rate of unit cost increases of any of the eleven countries in Table 6–4. When foreign costs are measured on a U.S. dollar basis (to account for relative changes in exchange rates), unit labor costs have actually declined substantially in Japan since 1978 (Capdevielle and Alvarez 1981).

TECHNICAL KNOWLEDGE, PRODUCTIVITY, AND CAPITAL

The "growth accounting" framework, first developed by Denison (1979) is often used to identify and measure the contributing sources of productivity growth. Denison estimates the contributions to economic growth made by changes in the quantity and composition of labor characteristics, changes in the quantity and composition of capital, resource allocation, changes in legal and environmental policies, economies of scale, and advances in knowledge. He comments that the division of productivity changes among some of their determinants "cannot be either complete or precise, but even partial and approximate estimates are useful if made with care" (1979: 64). The primary source of productivity growth, according both to Kendrick and Grossman and to Denison, is advances in knowledge. Denison concludes that "advances in knowledge is the biggest and most basic reason for the persistent long-term growth of output per unit of input." He explains the concept as follows (1979: 79):

> The term "advances in knowledge" must be construed comprehensively. It includes what is usually defined as technological knowledge—knowledge concerning physical properties of things and of how to make, combine, or use them in a physical sense. It also includes "managerial knowledge"—knowledge of business organizations and of managerial techniques construed in the broadest sense. Advances in knowledge comprise knowledge originating in this country and abroad, and knowledge obtained in any way: by organized research, by individual research, and by simple observation and experience.

Much of the knowledge that contributes to production—and to productivity—is actually embodied in physical capital—machines (hand tools, machine tools, computers) that give us the capacity to produce, both in the short and long run. As such, it is important to analyze "knowledge" in this context more deeply. Concep-

Table 6–4. Output Per Hour, Hourly Compensation, and Unit Labor Costs in Manufacturing, 1960–1980.

COUNTRY	1960-80	1960-70	1970-80	1960-73	1973-80	1977-78	1978-79	1979-80
OUTPUT PER HOUR								
UNITED STATES	2.7	2.9	2.4	3.0	1.7	0.9	1.1	-0.3
CANADA	3.8	4.3	2.7	4.5	1.9	1.6	1.7	-1.9
JAPAN	9.4	10.5	7.1	10.7	6.8	7.9	8.0	6.2
BELGIUM	7.2	6.3	7.0	7.0	6.2	6.0	5.8	3.6
DENMARK	6.4	6.1	6.1	6.4	5.1	4.4	2.3	1.7
FRANCE	5.6	6.1	5.0	6.0	4.9	5.3	5.4	0.6
GERMANY	5.4	5.7	5.2	5.5	4.8	3.8	6.3	-0.7
ITALY	5.9	7.1	4.6	6.9	3.6	2.9	7.3	6.7
NETHERLANDS	7.3	7.1	6.3	7.6	5.6	6.0	5.5	3.7
SWEDEN	5.2	6.9	2.9	6.7	2.1	4.3	8.1	0.6
UNITED KINGDOM	3.6	4.3	2.7	4.3	1.9	3.2	3.3	0.3
HOURLY COMPENSATION(2)								
UNITED STATES	6.7	4.5	8.8	5.0	9.3	8.2	9.8	10.7
CANADA	8.5	5.7	11.3	6.4	11.4	6.7	10.2	10.2
JAPAN	15.1	13.5	14.1	14.6	10.5	5.9	6.6	7.1
BELGIUM	12.4	9.7	14.1	10.7	12.0	7.0	7.4	10.1
DENMARK	13.3	11.0	14.7	11.8	13.1	10.3	10.9	10.7
FRANCE	11.8	8.7	15.3	19.3	15.2	12.9	13.9	15.0
GERMANY	10.3	8.6	10.9	9.4	9.7	8.5	9.1	7.9
ITALY	16.0	11.1	20.8	12.3	20.1	14.4	17.6	21.3
NETHERLANDS	13.2	12.0	12.9	12.8	10.6	8.6	7.5	6.9
SWEDEN	12.0	10.0	14.0	10.4	13.5	11.3	7.8	11.1
UNITED KINGDOM	12.7	7.2	18.5	8.6	19.3	16.5	19.3	23.6
UNIT LABOR COSTS: NATIONAL CURRENCY BASIS								
UNITED STATES	3.8	1.5	6.2	1.9	7.5	7.3	8.6	11.0
CANADA	4.5	1.3	8.4	1.8	9.4	5.0	8.3	12.4

JAPAN	5.3	2.7	6.5	3.5	3.4	−1.8	0.8
BELGIUM	4.9	3.2	6.6	3.5	5.5	1.0	6.3
DENMARK	6.5	4.6	8.1	5.1	7.6	5.6	8.9
FRANCE	5.9	2.4	9.8	3.1	9.9	7.3	14.3
GERMANY	4.7	2.7	5.5	3.7	4.7	4.6	8.7
ITALY	9.5	3.7	15.6	5.1	16.0	11.2	13.7
NETHERLANDS	5.5	4.6	6.2	4.8	4.8	2.5	3.1
SWEDEN	6.5	2.9	10.7	3.5	11.2	6.7	10.5
UNITED KINGDOM	8.8	2.8	15.4	4.1	17.2	12.8	23.3

UNIT LABOR COSTS: U.S. DOLLAR BASIS

UNITED STATES	3.8	1.5	6.2	1.9	7.5	7.3	11.0
CANADA	4.3	0.7	6.7	1.9	6.3	−2.1	12.6
JAPAN	8.0	2.8	11.8	4.9	8.3	26.2	−2.5
BELGIUM	7.8	3.2	12.5	4.6	10.6	15.1	6.7
DENMARK	7.9	3.6	11.6	5.0	9.3	15.1	1.8
FRANCE	6.5	1.7	12.4	2.8	10.9	17.1	15.3
GERMANY	9.3	3.4	13.2	6.1	11.2	21.0	9.8
ITALY	7.7	3.6	10.5	5.4	9.6	15.6	10.5
NETHERLANDS	8.9	4.8	12.8	6.1	10.3	16.4	4.2
SWEDEN	7.8	2.9	12.6	4.2	11.3	5.6	12.0
UNITED KINGDOM	6.6	1.0	12.9	2.6	15.3	24.0	27.7

(1) Rates of change computed from the least squares trend of the logarithms of the index numbers.

(2) Adjusted to include changes in employment taxes that are not compensation to employees, but are labor costs to employers (for France, Sweden, and the United Kingdom).

Note: These tables correspond with the article "International Comparisons of Trends in Productivity and Labor Costs", Monthly Labor Review, December 1981 plus subsequent revisions of the Canadian data released by statistics Canada on November 24, 1981. Data relate to all employed persons in the United States and Canada; all employees in the other countries.

Prepared By: U.S. Department of Labor, Bureau of Labor Statistics, Office of Productivity and Technology, December 1981.

tually, it is essential to distinguish between the quantity of capital in use and its quality, although it is operationally difficult to measure either the quantity or quality of capital in precise terms. If we assume that successive generations of new capital equipment are typically more and more productive because they embody an accumulation of knowledge, then the rate of accumulation of physical capital can be viewed as a combined measure of both the quantity and quality of capital. The importance of capital formation is emphasized by Fraumeni and Jorgenson (1980). They conclude that the contribution of capital input was the most important source of growth in aggregate value added during the period 1948-73 and that capital formation in the form of tangible assets dropped precipitously after the cyclical peak in economic activity in 1973.

There is a great deal of evidence documenting the slowdown in capital formation within the United States and the extent to which investment and savings within the United States lag behind other industrial countries. According to estimates given in Norsworthy, Harper, and Kunze (1979), the average annual growth rate of "business capital" in the total private business economy rose from 3.10 percent in the 1948-65 period, to 4.34 percent during 1965-73, and dropped to 2.23 percent in 1973-78.[2]

According to the American Productivity Center, the ratio of capital input to labor input grew at an average annual rate of 2.3 percent between 1965-73, falling to 1.5 percent in the 1973-80 period. In 1979, the ratio of gross fixed capital formation to gross domestic product was 17.5 percent in the United States, substantially smaller than in Japan (31.6 percent), Canada (22.9 percent), West Germany (21.3 percent), and several other of our trading partners (Sadler 1981: 10). The United States also has a low and declining personal savings rate, currently about 4.5 percent (down from 7.7 percent in 1975) compared to 20 percent in Japan, 14 percent in West Germany, and nearly 17 percent in France (Sadler 1981: 12). Bank savings per capita in Japan are more than

---

[2] They define "business capital" in the private business economy as net residential structures minus capital of nonprofit institutions plus tenant-occupied residential capital plus land and inventories. According to estimates given in Norsworthy (1982), the average annual growth of capital input in the U.S. manufacturing sector declined from 4.19 percent in the 1965-73 period to 3.06 percent during 1973-78.

twice as high as in the United States in absolute terms. Taking into account all sources of cash flow available for investment, the Japanese annually reinvest over 30 percent of their gross national product (GNP) compared to 21 percent for West Germany, 24 percent for France, and 18 percent or less for the United States. These figures understate the differential impact, since part of the annual investment must go to replace capital that is worn out or obsolete before there is anything left to support growth.

Available evidence suggests that the recent slowdown in capital formation within the private business economy has had some restraining effects on the growth rate of labor productivity. Norsworthy, Harper, and Kunze, of the Bureau of Labor Statistics, conclude (1979: 421):

> There are two distinct phases to the slowdown in the growth of labor productivity: 1965-73 and 1973-78. Differences are apparent both in the pattern of productivity growth among industries and in the factors contributing to the decline.
>
> The 1965-73 slowdown is largely unexplained by the factors we have considered. Capital formation was not a cause; changes in the composition of the labor force played a relatively minor role. Although R&D expenditures slowed during the period and may well have contributed to the productivity slowdown, we devised no satisfactory means to take this factor into account. Intersectoral shifts of capital and labor did not contribute.
>
> The 1973-78 slowdown is dominated by the effects of reduced capital formation. Some effect is also attributable to interindustry shifts in labor and capital. The sharp rise in energy prices may show up in a framework such as ours through its impact on capital formation and may help explain the relative weakness in capital formation in recent years.

The above paragraphs are based on their results, shown in Table 6–5. "Capital effects" accounted for over 70 percent of the slowdown in labor productivity in the private business sector of the economy during the 1973-78 period. Within the manufacturing sector, Norsworthy, Harper, and Kunze report that most of the slowdown in labor productivity could not be accounted for by either the capital effects or the labor effects included in their analysis.[3]

---

[3] The analysis described in Norsworthy, Harper, and Kunze (1979) is based on Gross National Product accounting methods where output is measured by "value added"—capital

Table 6–5.   Contributions to the Slowdown in the Growth of Labor Productivity in the Private Business Sector, 1965–78.

Item	1965–73 Slowdown	1973–78 Slowdown	Total
Change in labor productivity growth	−1.00	−1.12	−2.12
Contribution from capital effect			
Capital-labor ratio	−0.01	−0.54	−0.55
Asset composition	0.04	−0.05	−0.01
Intersectoral shifts	0.05	−0.14	−0.09
Pollution abatement capital	−0.03	−0.06	−0.09
Total	0.05	−0.79	−0.74
Contributions from labor effect			
Labor force composition	−0.06	0.04	−0.02
Interindustry shifts	0.05	−0.27	−0.22
Ratio of hours worked to hours paid	−0.09	0.05	−0.04
Total	−0.10	−0.18	−0.28
Contributions from effect of other factors	−0.95	−0.15	−1.10

The 1965–73 slowdown is measured relative to the 1948–65 period. The 1973–78 slow-down is measured relative to the 1973–76 period. Source: Norsworthy, Harper, and Kunze (1979: 416).

The quality of a capital good, such as a machine tool, is conceptually related to improvements in its capability to produce (e.g., finer tolerances, better control, better software, better interfaces with other machines). A classical problem in productivity analysis is to distinguish between the effects of more capital and those of "better" capital. The quality of capital is related to the knowledge embodied in it. Over successive generations, skills initially learned by

---

and labor costs, excluding materials purchases. This is an acceptable procedure when all sectors of the private business economy are aggregated together. However, when considering just the manufacturing sector, a value added measure of output distorts productivity measures because materials comprise such a large share (roughly 60 percent) of total production costs. More recently Norsworthy (1982) has shown that effects related to material inputs dominated the growth rate of labor productivity in the manufacturing sector both in the United States and in Japan between 1965 and 1977.

the workers are gradually shifted to the machines—permitting them to be operated by less skilled workers or to perform more specialized and intricate tasks. We also see some clerical and some managerial functions that can be shifted from humans to machines (i.e., computers). It is commonly assumed that capital goods produced each year embody the latest technological advances, so that the average age of capital goods would be a good indicator of capital quality and of the rate of diffusion of cost-reducing technology. According to Kendrick (1980: 26), throughout the U.S. economy the average age of capital declined by three years in 1946-66, by one year in 1973-77, and hardly at all during 1973-78.

Within the metalworking sector—those industries that manufacture machine tools, equipment, and all types of metal and electrical goods—we have direct evidence that the stock of machine tools is actually aging. According to the figures in Table 6–6, only 31 percent of the machine tools in the United States are less then ten years old, and 34 percent are over twenty years old. By contrast, in Japan 61 percent of machine tools are less than ten years old, and 18 percent are twenty years old or more.

It would be misleading to imply that increasing age (or decreasing relative efficiency) of the U.S. capital stock is *the* major determinant of economic performance. We merely note that U.S. productive capital stock is aging faster than that of other industrial countries, and that many products of the U.S. metalworking sector have recently been declining in international competitiveness. We have already mentioned the dramatic increases in import shares of televisions, semiconductors, autos, and trucks in Table 6–3. In recent years, trade balances in several categories of these and other goods have worsened, as shown in Figures 6–2 and 6–3). The six categories of goods, aside from chemicals, are products of the metalworking sector. The level of exports for these six products have declined since 1975, with light machinery, heavy machinery, and passenger cars experiencing the largest decreases.

Concluding Comments on Capital and Productivity

Most major studies of economic growth have acknowledged to varying degrees the contributing role of capital. "However, notwith-

Table 6–6.   Age Distribution of Machine Tools in Use.

	Percentage of Total Machine Tools in Use	Under 10 Years	10–19 Years	20 Years and Over	Percentage Numerically Controlled[a]
United States, 1976–78					
Primary metals	4.5	29.1	32.6	38.3	0.6
Fabricated metal products	24.0	27.4	35.2	37.4	0.9
Nonelectrical machinery	36.5	32.8	35.1	32.1	3.1
Electrical machinery	12.9	33.0	41.7	25.3	1.4
Transportation equipment	13.7	23.8	33.0	43.2	2.1
1. Motor vehicles	6.7	23.8	31.4	44.8	0.6
2. Aircraft and parts	5.3	23.4	67.7	41.6	5.3
Precision instruments	5.0	38.0	36.8	25.1	1.9
West Germany		37.0	37.0	26.0	NA
United Kingdom		39.0	37.0	24.0	NA
Japan		61.0	21.0	18.0	NA
France		37.0	33.0	30.0	NA
Italy		42.0	30.0	28.0	NA
Canada		47.0	35.0	18.0	NA

NA: not available.

[a] Machines which can operate automatically under control other than numerical control (e.g., mechanical cam control) are not included here.

Sources: American Machinist (1978) and National Machine Tool Builders Association (1981: 257–58).

standing all the studies that already have been undertaken," says productivity researcher Gary Fromm (1981: 110), "the precise role of capital formation in facilitating productivity improvement is not yet fully understood."[4] Even if our understanding of capital and economic growth were more complete, knowing that capital

---

[4] In the same paper, Fromm states, "A review of the macroeconomic evidence indicates that estimates derived by various investigators differ depending upon the form of the production functions utilized, the periods examined, the degree of aggregation and other characteristics of the data employed. It is also extremely hard to generalize and draw macroeconomic conclusions from microeconomic evidence because these studies normally involve different products, production processes and technologies "(1980: 110).

Figure 6–2.   Trade Trends for Light and Heavy Machinery and Its Manufactures.

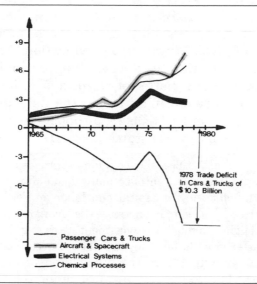

Passenger Cars & Trucks
Aircraft & Spacecraft
Electrical Systems
Chemical Processes

Source: Tesar (1980).

Figure 6–3.   Trade Trends for Other Dominant Categories of Manufacturers.

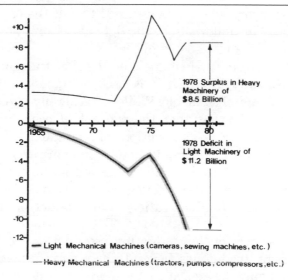

Light Mechanical Machines (cameras, sewing machines, etc.)

Heavy Mechanical Machines (tractors, pumps, compressors, etc.)

Source: Tesar (1980).

and so called "technical progress" in the aggregate have previously spurred general levels of economic growth may encourage a manager to invest but it cannot guide him in allocating resources among alternative types of machines or software or between physical capital, R&D, training, and education. At the micro economic level, at any rate, it is clear that continued productivity growth is not simply a matter of *more* capital. It is a question of the appropriate type of capital put in the right place. This point is emphasized by Von Furstenberg, although he is commenting on macroeconomic policies. He states (1980: 1):

> Effective remedial actions and coherent strategies will be difficult to devise institutionally, and intellectually demanding. They will have to stimulate the efficiency of capital and labor and the rate of technical change and not just simply increase the availability of inputs in the aggregate. Higher saving and investment rates may be necessary conditions for faster growth, but they are not sufficient conditions for maintaining that growth. Unless the distribution of investment is made more efficient and technology oriented, curing the "capital shortage" will not by itself provide a lasting cure for the "growth shortage."

Von Furstenburg's comments on the need to have more efficient capital in the right place requires careful consideration. Most macro- (and micro-) economic analysis is carried out in an "equilibrium" framework where it is assumed that if the capital is in place, "it works." This follows from the dubious assumption that previous decisonmaking and resource allocation has been made in an "optimal" fashion. Von Furstenburg is noting that the nature of the capital and the way in which it is used is equally important as the quantity of capital available. While we generally agree with with his comments, we suggest that the concept of economic efficiency be expanded to encompass the notions of adaptability over time, diversity, and rapid change, in addition to the narrowly defined concept of minimizing cost in a standardized and static environment.

We observe that international competition is forcing producers to increase both product variety and product quality simultaneously , while keeping down costs. These mutually incongruous requirements are pushing existing production technologies and management techniques beyond their current capabilities. This problem cannot be remedied by more capital investment per se. The solution requires balancing the needs for flexibility and efficiency within production and management systems.

LOW-, MID-, AND HIGH-VOLUME PRODUCTION

We now turn more specifically to the metalworking sector (SIC 33-38) and explain some of its important technological features.[5] The appropriate choice of capital and organization within particular establishments is strongly influenced by the average batch size—or the length of the average production run—in the factory. The more diverse the mix of parts (or products) being produced, the smaller the batch size for a given part. Some of the characteristics of the three "domains" of production—low-volume (or piece), mid-volume (or batch) and high-volume (or mass)—are shown in Tables 6–7, 6–8 and 6–9.

Table 6–7.   Overview of Discrete Parts Production.

| | Type of Process | | |
	Piece	Batch	Mass
Average total volume per part[a]	1-1,000	5,00-200,000	over 100,000
Average batch size[a]			
Small parts	1-100	100-5,000	Usually continuous
Large parts	1-10	10-5,000	Usually continuous or large batches
Estimated percentage of value added in metalworking sector[b]	10-20	60-80	10-30
Typical cutting time as percentage of available time[a]	4-7 (single shift) 12-21 (3 shifts)	10-15	20-38
Machining cost per part at full utilization	Highest	Low	Lowest

[a] Numbers vary with part size, part geometry, and manufacturing experience.

[b] Value added per unit produced is substantially higher in piece and batch production than it is in mass production. As a result, batch and piece production account for much of the value added within the metalworking industries, despite the fact that mass production accounts for most of the physical volume.

Sources: Adapted from Thomson (1980) and Borzcik (1980).

---

[5] Excellent overviews of the organization of production in the metalworking industries are given in Markowitz and Rowe, (1963b), Vietorisz (1969) and Westphal and Rhee (1983).

Table 6–8. Product-Process Overview of Piece, Batch, and Mass Production.

	Type of Process		
	Piece	Batch	Mass
Typical applications	Large aircraft; custom-designed machine tools; tools and dies; large turbines; space vehicles; maintenance and repair shops	Light aircraft; agricultural machinery; off-road vehicles; off-the-shelf machinery; trucks and buses; turbine blades; large electric motors;diesel engines; aircraft engines;	Cameras; watches; appliances; automobiles, transmissions and axles; auto engines; metal fasteners; bolts, nuts, and washers; metal drums and cans; lighting fixtures
Type of machine tools	Manual, or stand-alone NC	NC with some automated part handling, cells of machine tools, flexible manufacturing systems	Multistation rotary or transfer line, or specialized machine tools
Type of cutting tools	Largely standard tools	Some special tools	Mostly custom-made special tools
Automatic part handling	Very seldom	In many cases	Almost always
Likely size of company	Small shops; maintenance shops; speciality product shops	Any size	Mostly large companies

Sources: Adapted from Thomson (1980) and Borzcik (1980).

Table 6–9. Flexibility of Piece, Batch, and Mass Production Systems.

	Type of Process		
	Piece	Batch	Mass
Primary motivation	Ability	Flexibility	Volume and cost minimization
Part variety	Many different shapes and sizes	Mostly similar in shape, type of materials	Essentially identical parts with few processes, materials
Flexibility of machine tools	High—can combine several processes in one machine, and vary each process	Medium—limited to part family, can change tools, speeds, feeds, and dimensions	Low—usually only minor changes in speeds, feeds or part dimensions
Flexibility to change to a completely different part	Yes	Possible	Impossible
Flexibility to change to similar, but different part	Yes	Yes, if previously planned	Very limited
Flexibility to change materials	Yes	Limited	Extremely limited
Flexibility of plant layout	Smaller machine tools can be added	Some flexibility, but difficult to install new machines in system	Carefully planned, but not flexible; not usually possible to install new machines in system

Sources: Adapted from Thomson (1980) and Borzcik (1980).

There has always been a conflict between retaining a capability for rapid redesign or reconfiguration of the product, and achieving high levels of production and low unit cost. This is sometimes referred to as the flexibility versus efficiency tradeoff (See Chapter 3). For almost two hundred years, job shops have been considered flexible in the sense that manually operated metalworking machines, such as lathes, have been used to make a wide variety of different products on demand. But the unit costs in these labor-paced shops have necessarily been, and still are, considerably higher than the unit cost of products mass produced with specialized production equipment such as that employed in the auto industry. For example, according to Cook (1977), to machine a single automobile engine cylinder block with manually controlled, general-purpose machine tools would cost 100 times as much as to manufacture the same part by the most efficient mass-production methods.

In any batch production facility, equipment must be readily adaptable to make a new part within the "family" of parts being manufactured. The ease with which such changes can be made directly governs batch throughput time and costs. At present, it is expensive to change from one batch to another. This cost is the irreducible penalty of maintaining a desired degree of flexibility with respect to product change. By contrast, in mass production, the equipment is dedicated and optimized by design to specific operations at fixed rates and the plant achieves high efficiency by sacrificing flexibility. Flexibility and efficiency cannot be maximized simultaneously.

COST VERSUS BATCH SIZE

A schematic view of how unit manufacturing costs vary as a function of output with conventional technologies used in piece, batch, and mass production is shown in Figure 6–4. The quantity per period is measured on a logarithmic scale (10^1, 10^2, 10^3, etc.), and the vertical axis represents total (fixed and variable) production costs per unit of output. Typically, fixed costs go up and variable operating costs go down as the degree of automation increases. Some of the physical principles underlying these repre-

Figure 6–4. Average Cost Versus Batch Size.

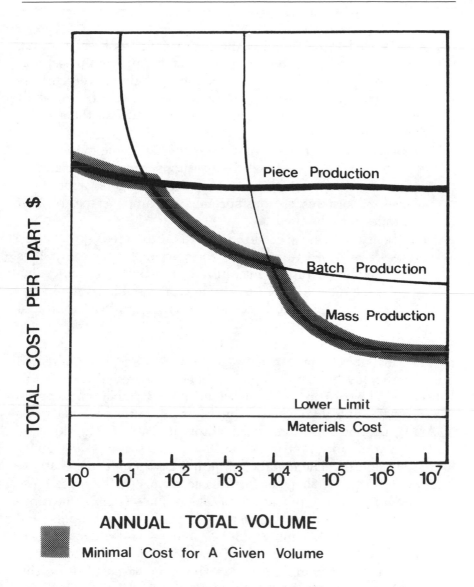

ANNUAL   TOTAL VOLUME

Minimal Cost for A Given Volume

Simplified curve resulting from a complex curve that includes the "step" changes of adding new machines, which are not fully utilized, to the overall cost factor of the product.

Source: Borzcik (1980: 65).

sentative average cost curves (and the cost-minimizing envelope) are well known to any production engineer:

- Small batch sizes require more frequent tool engineering, scheduling, materials handling, and setting up. Setup cost and other categories of fixed cost (including R&D and design) associated with each batch are spread over a small number of units, increasing the value added to each unit and the average cost per unit.
- As batch size increases, production flows become more nearly continuous. Cost associated with setup, support, and planning are spread over more units. As the batch size gets larger, more specialized systems, such as specialized clamps or fixtures and NC tapes may be used.
- For high volumes, specialized "hard automated" production processes are used to fabricate the part in a near continuous flow. Production rate is maximized, and unit cost and value added per unit are driven to the lowest possible levels. In the limit, total manufacturing cost tends to approach material cost.

The current relationship between unit cost and batch size can be seen roughly by comparing a number of representative metalworking industries in terms of value added (labor and capital costs plus profit margins) to the costs of purchased materials (VA/M), as shown for a group of sectors in Table 6–10. This VA/M ratio can be interpreted as the unit processing cost per dollar of materials used. The metalworking industries in Table 6–10 are similar in that they shape or form basic metals (mostly steel), use a minimum of highly processed metal products (such as bearings or engines) or of nonmetallic components as material inputs, and require little or no assembly.[6] Motor vehicles and other products largely assembled from purchased components are omitted from this table to avoid complications in the comparisons. Within this group, the highest VA/M ratio (almost 3.1) is for SIC 3544, special dies and tools, jigs and fixtures, the extreme case of custom or very small batch production. The lowest VA/M ratio (0.68) is for

[6] Some of the products made within industry SIC 3544, special tools, dies, and jigs and fixtures, do use highly processed material inputs and require some assembly.

Table 6–10.   Unit Cost Index of Selected Steel Products.

Product	SIC Code	1977 Value Added ($ times 10⁶)	Materials Cost ($ times 10⁶)	Cost Ratio
Special dies, tools, jigs and fixtures	3544	2,790.1	907.3	3.07
Cutlery	3421	492.0	172.8	2.85
Hand and edge tools	3423	1,421.4	642.0	2.21
Carburetors, pistons, rings, and valves[b]	3592	924.9	451.9	2.05
Screw machine products	3451	1,023.9	619.2	1.65
Bolts, nuts, washers, and rivets	3452	1,840.3	1,229.9	1.50
Ball and roller bearings	3562	1,472.7	1,018.6	1.44
Iron and steel forgings	3462	1,301.4	1,297.7	1.00
Automotive stampings	3465	4,654.5	4,862.3	0.96
Metal barrels, pails, etc.	3412	389.6	528.6	0.73
Metal cans	3411	3,159.3	4,643.4	0.68

[a] Materials cost includes all raw materials, semifinished goods, parts, containers, scrap, and supplies and excludes fuels, electricity, resales, and material used for subcontracted work.

[b] Includes establishments not classified in the motor vehicles industry.

Source: Census of Manufacturers (1981: Table 3a).

SIC 3411, metal cans, an extreme case of mass production. The range between these two extremes is almost five. That is, SIC 3544 spends nearly five times as much in capital and labor cost for each dollar of material processed as SIC 3411.[7] In general, batch size increases as we go down the list from highest to lowest VA/M ratios.

The true range in processing cost ratios is actually much greater when differences in the unit cost of inputs to the different industries are considered. Those industries with the lowest batch size also, in general, use high-cost alloys. The mass production

---

[7] Some of the unit cost differential between these two industries is due to differences in the amount of processing required to change the material from its original shape to the shape of the final product, which is unrelated to differences in batch size. Rolled tin or aluminum is already very close to the shape of a can whereas a bar of steel often requires a severe geometric change to be transformed into a special tool.

industries typically use lower grades of materials, such as ordinary carbon steel.[8] When the differences in material inputs is taken into consideration (e.g., a ton of high-speed tool steel alloy costs nearly fifteen times more than a ton of carbon steel), sector SIC 3544 probably adds more than ten times as much "value" per pound of material processed as sector 3411. As a rough working estimate, unit costs are at least ten times higher for custom production in a job shop environment as they are for mass production operations. In the intermediate batch processes, unit costs range from three to five times higher per pound of material processed than in the most extreme mass production case, when adjusted for the relative cost of different qualities of materials.

Clearly, producers and consumers are paying a high premium for metal products produced in small batches. According to some available estimates, between 70 and 90 percent of the value added in the metalworking sector is attributable to batch and piece production (Table 6–7). This suggests that much of the value added in the metalworking sector can be thought of as the cost of flexibility needed for product differentiation and specialization.

NARROWING THE GAP BETWEEN BATCH AND MASS PRODUCTION

The unit cost ratios provide a basis for estimating how much of a cost saving might be realized by making small- and medium-volume production closer in efficiency to current mass production processes. It is not unreasonable to suppose, for purposes of argument, that robotics and CAM could eventually reduce the differential in unit processing cost between batch and mass production

---

[8] The ratio of value added to materials cost must be adjusted to reflect differences in the proportions of value added "embodied" within the materials purchased by the various industries. These differences distort interindustry comparisons. For example, in SIC 3544, alloy and stainless steels, which are more expensive than carbon steel, account for nearly 10 percent of material cost, and processed components, such as bearings, stampings, motors, and speed changers account for almost another 5 percent. A more complicated product, such as a machine tool or an automobile, uses an even larger proportion of highly processed material inputs, which would further complicate comparisons. To rectify this problem, the heterogeneous material inputs for each industry must be "purified" to a standard bill of materials that can be meaningfully compared across industries. One workable procedure is to compare value added per ton of metal processed in the industry.

by a factor of two or more. To estimate the level of savings that could be achieved, the value added for the industries listed in Table 6–10 is recomputed, assuming that they process the same value of materials but that the current difference between the VA/M ratio for the industry and the minimum value of VA/M shown in the table (metal cans, VA/M = 0.68) is cut in half. This is probably a conservative estimate of the potential. Table 6–11 shows the dollar savings and the percentage price reduction (assuming price reductions are passed on) that would be realized on this basis.[9] Implications for more complex products involving assembly, such as automobiles and appliances, would be qualitatively similar. Robots are already playing a major role in reducing the costs of welding, spray painting, and inspection in auto manufacturing. The U.S. auto manufacturers will need to reduce their processing costs (capital and labor costs) by 30 percent or more during the next decade or so merely to equal the present efficiency of Japanese auto manufacturers. Cheaper components and capital equipment would obviously be very welcome in Detroit.

This type of analysis is useful only as a first-order approximation of cost savings, since it assumes that unit costs in batch (and custom) production will approach the lower levels of unit costs in mass production, without specifying how this will happen. A more detailed examination of how much the major components of unit cost—labor cost, the level of output, capital cost, and material cost—might be altered and of how product quality might be improved contributes to a better understanding of the potential impacts of robotics on reducing unit cost and on increasing manufacturing productivity. We shift our emphasis here, ignoring for the moment the important distinctions previously made between piece, batch, and mass production processes, and focus on the the components of unit cost and of quality within the metalworking sector as a whole. The main components of unit costs are listed in Table 6–12, along with key variables that may be used to analyze the impact of robotic technology on a given cost component. Several important aspects of product quality are also listed because of their great importance to competitivenenss. The potential impacts on labor cost and on capacity are considered in greater de-

---

[9] The current average unit cost in mass production industries is not a lower bound on cost levels, since programmable automation is also being installed in mass production plants.

Table 6–11.   Potential Cost Savings by Eliminating the Batch-Mass Gap.

Product	Potential Savings ($ × 10⁶)	Effective Price Reduction (%)
Special dies, tools, etc.	1089	28%
Cutlery	187	26%
Hand and edge tools	494	22%
Carburetors, pistons, etc.	308	22%
Screw machine products	302	17%
Bolts, nuts, etc.	500	15%
Ball and roller bearings	393	15%
Iron and steel forgings	211	8%
Automobile stampings	667	7%
Metal barrels, pails, etc.	17	2%
Metal cans	—	—

Source: Derived from data in Table 6–10.

Table 6–12.   Estimating the Impacts of Robotics Technology on the Components of Unit Cost.

Cost Component	Key Variables for Estimating Changes
Labor cost	Percentage of workers, by industry, by occupation, and by skill level that could be replaced by level I (nonsensor-based) and level II (sensor-based) robots
Capital cost	Percentage change in inventory cost
	Percentage change in annual capital cost resulting from purchasing robots
Material cost	Percentage change in unit material requirements due to reduced wastage (failure to meet specifications)
Capacity	Percentage change in days per year and shifts per day scheduled for production
	Percentage change in throughput per unit time the equipment is in operation
Quality Concerns	
	Ability to consistently meet existing performance standards
	Ability to exceed existing performance standards
	Ability to build products which can provide new services.

tail. Potential impacts on capital are only mentioned briefly, and changes in material costs are not discussed here.[10]

As previously discussed in Chapter 3, robotic technology also enables producers to improve product quality. Robots can perform many repetitive tasks, such as painting or welding, more consistently and more accurately than people, making it possible to meet exisiting performance standards more consistently, and even meet more stringent standards. "Flexible" production technologies which can be reprogrammed, such as robotics, make it less expensive to alter existing product designs. This capability, in conjunction with more consistent and more accurate control of operations, makes it economical and technically feasible to experiment with, and produce designs which otherwise could not be manufactured. These quality factors, which enhance the value and the competitiveness of the product, are not dealt with further, even though they might ultimately be of more importance than reductions in the cost of producing the current mix of goods and services.

POTENTIAL IMPACTS ON LABOR COST

Existing and likely near-term capabilities of robots make them candidates to replace significant numbers of machine operators and unskilled laborers over the next three decades or so. In general terms, the replacement of several million production workers by robots will decrease direct labor input. Consequently, output per labor hour will most certainly increase. Total factor productivity also depends on the requirements for new capital vis à vis utilization of existing capital and on other ways in which robotic (and CAD/CAM) technology may alter productive efficiency and capacity.

To gauge roughly how much of an impact the direct substitution of robots for factory workers could have on labor cost, all other factors constant, we compare 1977 labor payments in the metalworking sector and in all manufacturing with a reduced total derived from our previous assumptions about the potential for robotic substitution discussed in Chapter 5. The value of ship-

---

[10] See the discussion of material savings in Chapter 3.

ments for 1977 is decomposed into labor cost, materials cost, and other costs (capital costs, inventory adjustments, and profits) in Figures 6–5 and 6–6. Wage and benefit payments to production workers accounted for 17 percent of the value of shipments in metalworking and for 12 percent throughout all manufacturing. This breakdown of cost proportions understates the total labor cost of metalworking products, since purchased materials account for the largest share of the value of output, and most of the material inputs originate from industries within the metalworking sector. The "true" labor content is the accumulated sum of labor payments made within an industry, as well as the labor cost "embodied" in the purchased materials.[11] For now, we consider an industry (or sector) in isolation, without considering the effects of cost savings that might be passed on to other industries.

The three occupational groups most amenable to robotization—metalworking craft workers, semiskilled operatives, and laborers—comprise almost three-quarters of total production workers. Potential reductions in labor payments are calculated by assuming that 16 percent of the workers in these three occupational groups could be replaced by level I robots, and 40 percent by level II robots (See Chapter 5). The cost reductions in metalworking (without considering interindustry interactions) appear to be quite modest, ranging from 2.1 percent (level I substitution rate) to 5.2 percent (level II substitution rate). For all manufacturing,

---

[11] The motor vehicle and car body industry (SIC 3711), where cars are assembled, provides an extreme example of the extent to which labor payments made within the industry alone can understate the total labor cost of the product. In this industry, roughly twelve cents of every dollar earned go toward total wage and salary payments. General Motors, a conglomerate encompassing all aspects of auto production including the manufacture of parts, engines, transmissions, as well as final assembly, pays approximately thirty cents on the dollar for wage and salary payments. The explanation of the difference is that plants that supply the various car components—engines, transmissions, stampings, interior trim, and the like—are classified as separate industries (most notably, motor vehicle parts and accessories, SIC 3714 and automotive stampings, SIC 3465), even though they are often owned by the same company. As a result, over 75 percent of the cost in auto assembly plants goes toward materials purchased from plants classified under other SIC codes, with over two-thirds of the total labor content of the car embodied in the materials.

The notion of accumulated value added (including labor cost) is more easily grasped by examining how many dollars of product from industry i and how much value added (employee compensation and capital expenses) are required to produce a dollar's worth of output in industry j. This number, know as the input coefficient, is derived from the matrix of interindustry transactions compiled by the Department of Commerce. The notion of accumulated value added is discussed further in Stone (1962) and Ayres (1978).

Figure 6–5. Average Cost Proportions in Metalworking (SIC 33-38), 1977.

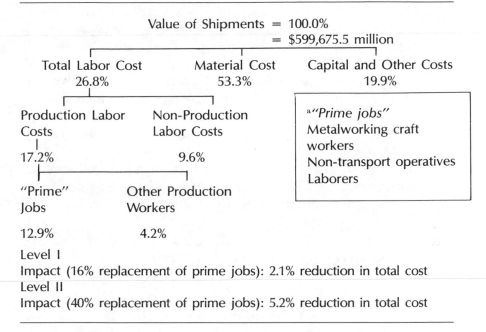

Value of Shipments = 100.0%
= $599,675.5 million

| Total Labor Cost 26.8% | Material Cost 53.3% | Capital and Other Costs 19.9% |

| Production Labor Costs 17.2% | Non-Production Labor Costs 9.6% |

[a]*"Prime jobs"*
Metalworking craft workers
Non-transport operatives
Laborers

| "Prime" Jobs 12.9% | Other Production Workers 4.2% |

Level I
Impact (16% replacement of prime jobs): 2.1% reduction in total cost
Level II
Impact (40% replacement of prime jobs): 5.2% reduction in total cost

Figure 6–6. Average Cost Proportions in All Manufacturing (SIC 20-39), 1977.

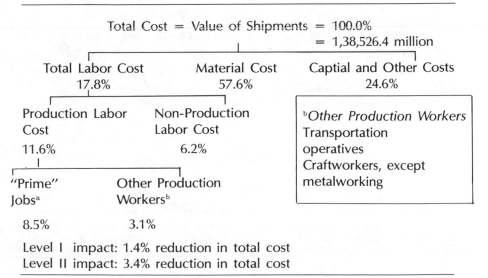

Total Cost = Value of Shipments = 100.0%
= 1,38,526.4 million

| Total Labor Cost 17.8% | Material Cost 57.6% | Captial and Other Costs 24.6% |

| Production Labor Cost 11.6% | Non-Production Labor Cost 6.2% |

[b]*Other Production Workers*
Transportation operatives
Craftworkers, except metalworking

| "Prime" Jobs[a] 8.5% | Other Production Workers[b] 3.1% |

Level I impact: 1.4% reduction in total cost
Level II impact: 3.4% reduction in total cost

Source: Derived from data in Census of Manufacturers (1981).

potential cost reductions are somewhat lower, ranging from 1.4 percent (level I) to 3.4 percent (level II).

Under the highly restrictive assumptions that the quantity and mix of physical output remain at current levels and that the organization of production remains unchanged (both unlikely assumptions), the direct substitution of robots for workers would eventually increase labor productivity 24 percent (level II impacts). If this were to happen over a twenty-year period, the annual average improvement factor would be 1.1 percent. Thus, it does not appear that the direct substitution of robots for factory workers by itself would have a substantial impact on total cost or labor productivity throughout manufacturing, or even within metalworking, where robot use is most heavily concentrated. Something else must happen if large increases in productivity are to be realized.

POTENTIAL IMPACTS ON CAPACITY

Possibly more significant is the impact of robotization and CAM on factory organization and the utilization of capital. There are two types of capital that robotics and CAM can help to utilize more effectively: machines and working capital tied up in inventory.

In the metalworking sector (SIC 33-38), the effective utilization of manually operated metal cutting machines, metal forming machines, and welding equipment is remarkably low. According to our estimates in Table 6–13, the figures are 12 percent, 15 percent, and 22 percent respectively, assuming "utilization" corresponds to twenty hours per day, seven days a week, with an operator present. These estimates only measure the proportion of time an operator is available to run the machine, with no consideration of whether the machine is in use when the operator is on duty.[12] According to estimates shown in Figure 6–7, productive

---

[12] These figures are only crude estimates of utilization. Most factories do not schedule preventitive maintenance on a daily basis, so theoretical utilization could reasonably be calculated assuming twenty-four available hours per day. This would result in lower estimates of machine tool utilization. On the other hand, the estimates might be biased on the low side if we consider that older, fully depreciated machine tools are often left idle intentionally for surge capacity, or for special types of jobs. To purify the estimates, we might give more attention to the utilization of the newer automatic and numerically controlled machines than to the utilization of the older, manually controlled machines.

Table 6–13.   Estimates of Average Machine Tool Utilization in the Metalworking Industries, 1977.

Sector	Metal Cutting Tools (%)	Metal Forming Tools (%)	Joining (Welding) Tools (%)
33	17.8	35.5	24.4
34	11.1	15.6	17.2
35	11.4	9.6	21.8
36	8.6	14.3	10.2
37	15.3	20.3	40.6
38	7.3	6.9	13.6
Average	12	15	22

Assumptions: Full utilization of a stand alone, manually controlled machine tool would be equivalent to 2.5 shifts (20 hours/day) operation, seven days a week. This corresponds to 7,280 man-hours per year. Assume 2,000 hours per worker per year. Thus one manually controlled tool requires 3.6 operators per machine per day.

$$\text{Utilization} = \frac{\text{Number of non NC machine operators} * 2,000}{\text{Number of non NC machines} * 7,280}$$

Sources: Number of machine tools: American Machinist (1978); number of operators: Bureau of Labor Statistics (1980b).

cutting time by machine tools as a fraction of theoretical capacity in low- and medium-volume shops is 6 percent and 8 percent respectively, increasing to 22 percent for high-volume, mass-production operations. Thus, it is clear that most machines, especially in the metalworking industries, are idle most of the time—even before making allowance for setup time, load/unload time, and other adjustments. Even in the most automated industries, such as those which produce automobile engines and transmissions, true machine tool utilization rates as high as 50 percent are seldom, if ever, achieved at present. We believe the major quantifiable economic impact of robotics and CAM will be to expand sharply the effective capacity of production facilities by increas-

Figure 6–7.   Breakdown of Theoretical Capacity in Low-, Mid-, and High-Volume Manufacturing.

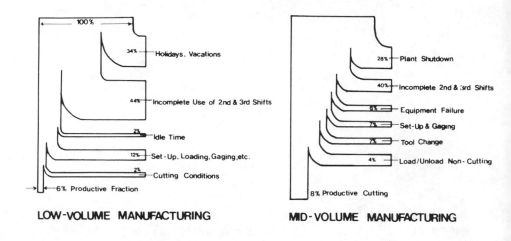

LOW-VOLUME MANUFACTURING

MID-VOLUME MANUFACTURING

HIGH-VOLUME MANUFACTURING

Source: Mayer and Lee (1980a: 32)

ing both the amount of time per year the plant is operating and the throughput per shift. Much of the lost time is due to incomplete use of the second and third shifts and to plant closings for holidays, strikes, and other reasons. Weekends, holidays, and night shifts are less popular than the "normal" forty-hour work week. Consequently, even if labor is available during these periods, it is more expensive.

Average estimates of the days per year plants are open and of the shifts per day they are actually operating are given in Table 6–14. We deduced these estimates from the breakdowns of theoretical capacity shown shown in Figure 6–7. These are intended to represent normal operating conditions in a "healthy" econo-

Table 6–14. Estimates of Planned Production Time in Low-, Mid-, and High-Volume Metal Fabricating Manufacturing.

	High Volume	Mid Volume	Low Volume (1 shift operation)	Low Volume (2 shift operation)
Maximum days per year available	365	365	365	365
Days per year open for operation	286	263	241	241
Hours per day scheduled for:				
Production	22.2	10.7	8.0	16.0
Preventive maintenance[a]	1.0	1.0	1.0	1.0
Not scheduled (idle)	0.8	12.3	15.0	7.0
Scheduled production time as a fraction of maximum available time	72.6%	32.0%	22.0%	44.0%

[a] Most factories do not stop production on a daily basis to perfom scheduled preventive maintenance. Machines are typically serviced on an "as needed" basis. Assuming 1 hour per day of scheduled maintenance may even be a high estimate. Major machine overhauls and repairs are typically carried out during scheduled plant shutdowns.

Source: Derived from breakdowns of theoretical capacity in Figure 6–7.

my.[13] These figures imply that high-volume plants shut down nearly 80 days per year due to Sundays, holidays, and planned closings for retooling. Mid-volume plants are closed, on average, 102 days (all weekends), and low-volume plants are closed nearly 125 days out of the year (weekends plus three weeks for holidays and shutdown). When open for production, high-volume plants are typically operating over 22 hours per day, whereas mid-volume and low-volume plants are typically scheduled to operate 10.7 hours and 8 hours per day, respectively. Clearly, there is considerable potential for increasing output (and thereby decreasing unit cost) by saving long runs for an unmanned third shift (or weekend), using robot operators. During the next two decades, as manufacturers gain experience with unmanned factory operations, the less routine machine setups, repair, maintenance, and inspection tasks could be reserved for the regular day shift. Planned shutdowns required for retooling would be substantially reduced, or possibly eliminated in a robot-integrated factory with flexible production technologies. A summary of the potential for increasing available production time (and hence output) in existing facilities is shown in Table 6–15. Output per year could be increased by perhaps 30 percent in high-volume plants and by almost 200 percent in mid-volume plants. If low-volume plants are operating only one shift per day, as is suggested by the breakdowns in Figure 6–7, recouping lost time could increase output by over 330 percent! If we generously assume that low-volume shops are operating on a 2 shift basis (which only a portion do), output per year could still be increased by nearly 200 percent.

In addition to extending the amount of working time per year, robotics, especially when integrated with other CAM technologies, can increase capacity by increasing the number of parts produced per hour. Based on the breakdowns of theoretical capacity in Figure 6–7, machines are productively engaged only about 30 percent of the time that the plant is scheduled for production. The remaining 70 percent of planned production time is lost for a varie-

---

[13] Actual figures vary with demand. The high volume manufacturing estimates are based on operations of several machining plants for an automobile producer before the sharp cutbacks in production. According to a survey of plant capacity conducted by the Bureau of Census (1980), most metalworking industries are only producing 60 to 80 percent of their potential output, that lack of sufficient orders was by far the most important reason that operations fell short of practical capacity.

Table 6–15. Potential Percentage Increases in Output from Utilizing Lost Time.

Type of Plant	From Utilizing Days Plant Is Closed	From Utilizing Nonscheduled Production Time	Total Percentage Increase in Output
High-volume	28	3	31
Mid-volume	83	115	198
Low-volume (1-shift operation)	148	187	335
Low-volume (2-shift operation)	148	25	192

Source: Derived from estimates of available time in Table 6–14.

ty of reasons. Recouping the fraction of planned production time that is "lost" to nonproductive uses would further increase the effective capacity of machine tools.

Much of the "lost" time is machine related—equipment limitations, tool changing and equipment failures. However, a sizable fraction of time lost is due to management and workforce practices, including personal time breaks, late starts and early quits, material handling, excessive machine adjustments, and in-line storage losses due to scheduling inefficiencies. Personal time, late starts and early quits, and some fraction of the material handling time could be expected to be nearly eliminated by replacing workers with robots. Time losses due to tool changing, equipment failures, excessive machine adjustments, setups, and scheduling inefficiencies will probably not be affected directly by robots but might be reduced if more aspects of factory work were consolidated and controlled by sensor-based computer systems. For example, sensors monitoring machine performance would eliminate unnecessary adjustments and would speed up diagnosis of machine failures.[14] If

---

[14] In the next few years, time lost to equipment failures could conceivably increase as systems become more automated and more complex. However, we expect improvements in machine reliability and in sensor-based diagnostic systems to improve machine and system reliabilty and to reduce equipment failures over the next two decades.

"stand-alone" machines were replaced by a flexible manufacturing system, and parts processing was "rationalized" by adopting group technology, there would be less material handling and the scheduling of parts and tools would be simplified. Even a substantial fraction of the equipment related losses could be eliminated in a fully integrated flexible manufacturing system, since the whole system need not be stopped if one station malfunctions. Robots or programmable pallets under the control of a central scheduling computer could reroute parts to other work stations.

It is difficult to discuss the potential improvements in productivity that may be brought about from robotics in isolation of the development of CAM systems and other forms of factory automation. Retrofitting robots into exisiting production lines would bring about some improvements, such as increasing the utilization of a single machine or work station, but we do not expect that it would dramatically improve overall factory performance. Substantial impacts on performance and cost at the factory level require the integration of robots and other forms of factory automation into coordinated manufacturing systems. Also, it becomes more difficult and less meaningful to distinguish between robots and other forms of factory automation as the concept of robotics evolves from programmable manipulators to machines and systems which can "sense, think and act."[15]

Our own rough estimates of potential increases in throughput that could be achieved from recouping the "nonproductive" time lost during scheduled operations are shown in tables 6–16 through 6–18. These estimates are based only on informed judgment but have been reviewed by several industry experts. They are not the result of detailed analysis. We distinguish two levels of improvement: 1) as a result of introducing robots only and 2) as a result of integrating robots with CAM systems and other forms of factory automation. We suggest that the installation of robots, without increasing the time normally planned for operations and without extensively adding other forms of automation, would result in a 10 percent increase in output in high-volume machining operations (not including assembly) and nearly a 15 percent increase in output in mid-volume and low-volume produc-

---

[15] We borrow the broader definition of robotics as machines which can "sense, think and act" from Professor Raj Reddy, Director of the Carnegie-Mellon Robotics Institttue.

Table 6–16.   Potential Percentage Output Increases from Recouping Non-Productive Time during Planned Operations: High Volume Manufacturing.

Function	Percentage of Operating Time	Potential Percentage Reduction (Robots Only)	Adjusted Percentage	Potential Percentage Reduction (Robots with CAM)	Adjusted Percentage
Load/unload, noncutting[a]	20	−10	18	−25	15
Workstation allowances	20	−40	12	−-80	4
Inadequate storage	10	0	10	−50	5
Tool change[a]	10	0	10	−20	8
Equipment failure	10	0	10	0	10
Productive fraction	3	0	30	0	30
Total[b]	100.0		90.0		72.0
Potential output index	1.00		1.11		1.39

[a] Already highly automated in high volume plants.
[b] Total equals total scheduled production time.
Sources: Breakdown of theoretical capacity: Dalles (1981) and Mayer and Lee (1980b). Estimates of potential improvements: Ayres and Miller.

tion. If robots were used in conjunction with other forms of factory automation systems, still without increasing the number of days normally planned for operations, output might be increased by nearly 50 percent in mid- and low-volume production and possibly by 40 percent in high-volume production. Assuming there is a market for additional goods, an increase in output would result in a reduction in unit cost. For comparison purposes, in our discussion of the impacts on labor costs earlier in this chapter, we noted that if level II (sensor-based) robots displaced 40 percent of the production workers whose jobs could potentially be replaced by automated systems over the next two decades, total costs within the metalworking industries would be reduced by roughly

Table 6–17. Potential Output Increases from Recouping Non-Productive Time during Planned Operations: Mid-Volume Manufacturing.

Function	Percentage of Operating Time	Potential Percentage Reduction (Robots Only)	Adjusted Percentage	Potential Percentage Reduction (Robots with CAM)	Adjusted Percentage
Setup and gauging	22	−30	15.4	−65	7.7
Load/unload and noncutting	12	−40	7.2	−60	4.2
Tool change	22	−5	20.9	−15	18.7
Equipment failure	7	0	7	0	7
Idle time	12	0	12	−25	9
Productive fraction	25	0	25	0	25
Total[a]	100		87.5		64.6
Output index	1.0		1.14		1.55

[a] Total equals total scheduled production time.
Sources: Breakdowns of theoretical capacity from Figure 6–7. Estimates of potential reductions: Ayres and Miller (1981).

5 percent. If all of these production workers were replaced with robots and other types of machines, total costs would be decreased by approximately 13 percent.

Our judgments regarding the estimated improvements in Tables 6–16 through 6–18 are mostly applicable to situations of "retrofitting" or incrementally adding technologies within existing plants. Other studies suggest that completely redesigning the factory around new technology could result in substantially greater improvements in throughput during planned production periods. For example, Mayer and Lee (1980b), of Ford Motor Company, estimated the combined effects of applying the most advanced concepts to almost all aspects of machine design and control and factory layout in high-volume machining systems. They considered the use of automatic loading/unloading, automatic tool changing, diagnostic sensing, component reliability, and un-

Table 6–18.   Potential Output Increases from Recouping Non-Productive Time during Planned Operations: Low-Volume Manufacturing.

Function	Percentage of Operating Time	Potential Percentage Reduction (Robots Only)	Potential Adjusted Percentage	Potential Percentage Reduction (Robots with CAM)	Potential Adjusted Percentage
Setup, loading, and gauging	55	−25	41.3	−50	27.5
Idle time	9	0	9	−50	4.5
Cutting conditions	9	0	9	−25	6.8
Productive fraction	27	0	27	0	27
Total[a]	100		86.3		65.8
Output index	1.0		1.16		1.52

[a] Total equals total scheduled production time.
Sources: Breakdowns of theoretical capacity from Figure 6–7. Estimates of potential reductions: Ayres and Miller (1981).

manned operation, as well as improved line balancing and faster cutting speeds. Their results suggest that all of these improvements, *without* increasing cutting speeds, could result in a 90 percent increase in output. If cutting speeds were increased to their upper limits as well, over a 200 percent increase in output could be achieved. Their estimates are also based on *current* operating times (no changes in shifts per day or days per year).

The potential effects of utilizing nonscheduled production time and of recouping time lost to nonproductive uses during scheduled operations are combined in Table 6–19. To utilize all of the time normally not scheduled for production (plant shutdowns, holidays, Sundays) the plant would sometimes have to operate with "skeleton" crews, or even unmanned during some periods, under the control of computer systems. Thus, if we assume that hours available for production could be increased to its upper limit, we essentially require that we consider the case of robots used in conjunction with other CAM technologies. For the "robots with

Table 6–19.   Summary of Potential Impacts on Capacity.

Type of Plant	Base Case	Potential Capacity Increases	
		Robots Only	Robots with CAM
High volume			
Available hour index	1.00	1.31	1.31
Throughput index	1.00	1.11	1.39
Output index	1.00	1.45	1.82
Percentage increase in output		**45**	**82**
Mid-volume			
Available hour index	1.00	2.98	2.98
Throughput index	1.00	1.14	1.55
Output index	1.00	3.40	4.62
Percentage increase in output		**240**	**362**
Low-volume: single shift			
Available hour index	1.00	4.35	4.35
Throughput index	1.00	1.16	1.52
Output index	1.00	5.05	6.61
Percentage increase in output		**405**	**561**
Low-volume: double shift			
Available hour index	1.00	2.92	2.92
Throughput index	1.00	1.16	1.52
Output index	1.00	3.39	4.44
Percentage increase in output		**239**	**344**

Sources: Derived from data in Tables 6–15, 6–16, 6–17 and 6–18.

CAM" case, high-volume machining operations show a potential output increase of 80 percent. Mid-volume manufacturing, and low-volume producers already on a double shift show a potential output increase of 350 percent. For low-volume producers operating on normal single shifts, the potential increase is 560 percent. In mid- and low-volume manufacturing, potential increases in output are almost all the result of increasing planned production time. In high-volume machining opertions, the contribution of increasing throughput per period is somewhat greater than the contribution of increasing hours planned for production. *Potential cost savings theoretically obtainable from increasing machine utilization appear to be order of magnitude larger than savings which could be realized be replacing direct production labor.*

POTENTIAL IMPACTS ON CAPITAL COSTS

Inventory and Work-In-Process

It is estimated that in typical batch production operations, 95 percent of the time a workpiece is in the factory, it is in transit or storage.[16] In other words, the piece is being actively worked on only 5 percent of the time. This suggests that a piece requiring 10 hours of machining time would take at least 8 days to pass through the factory. Because of backorders and the complexities encountered in assigning parts to machines, pieces requiring 10 hours of machining sometimes take several weeks or longer to pass through many batch production factories. Improved work scheduling, in combination with higher rates of machine utilization could dramatically reduce the time required to move work through the factory.

A rough indication of the magnitude of capital tied up in inventory can be seen from Table 6–20 which shows end-of-year inventories as a percentage of the yearly shipments for the industries in our sample. These figures include inventories of finished goods, work-in-process and materials, supplies, and fuel. Robotics and flexible manufacturing systems will not necessarily decrease the levels of inventory on hand. In fact, they may require increased inventory levels, since higher levels of machine utilization will mean higher output rates, and it will be more important to keep a "buffer" supply of work always available to keep the machines busy. However, the financial benefits of reducing the time it takes to move inventory through the plant can be seen from the following example. Suppose a shop with an average of $1 million annual revenues has a 100-day average lead time, which is not atypical. If that lag could be "instantly" cut to 10 days, the firm would be able to put the next 90 days' revenue ($250,000) into the bank, without increasing outlays at all. This amount would effectively be converted from unavailable working

---

[16] See Carter 1972), and Eversheim 1980).

Table 6–20.   Inventory as a Fraction of Value of Shipments.

Industry	1977 Value of Shipments ($ × 10⁶)	Inventory Ratio (%)
Special dies, tools, etc.	3,909.6	11%
Cutlery	711.1	18%
Hand and edge tools	2,220.4	21%
Carburators, pistons, etc.	1,362.4	17%
Screw machine products	1,759.0	13%
Bolts, nuts, etc.	3,311.0	21%
Ball and roller bearings	2,575.3	24%
Iron and steel forgings	2,806.8	16%
Automobile stampings	9,708.7	9%
Metal barrels, pails, etc.	461.3	29%
Metal cans	8,097.7	12%

Inventory includes finished goods, work-in-process, materials, suppliers, fuels, and other inventories. Inventory ratio = end of year inventories/value of shipments. Census procedures make no adjustments for the inventory valuation procedures used by respondents.
Source: Derived from data in Census of Manufacturers (1981).

capital into cash. The firm would also save significantly on warehousing and other inventory-related costs.

While a 90 percent reduction in lead time cannot be expected across the board (especially in mass production operations), it is well within the range of possibilities in some capital goods industries, including the aerospace and machine tool producers. Edward Nilson of Pratt and Whitney Aircraft, a major producer of jet engines, was recently quoted in *Fortune* (Bylinsky 1981): "Thanks to CAD we have gained a 5-to-1 or 6-to-1 reduction in labor and at least 2-to-1 reduction in lead time. And these ratios go up as high to 30-to-1 and 50-to-1 when we link CAD to CAM." If machine tool manufacturers could cut their lead time from two years or more to a few months, the automobile manufacturers, in turn, could cut the lag in bringing out new models by perhaps 18 months.[17] If this improved turnaround capability had been avail-

---

[17] The automobile planning cycle may stretch back as far as 120 months prior to vehicle introduction, if concept research is included. Once a primary program is selected, there is

able six years ago, the Japanese might not have established such a strong foothold in the U.S. auto market.

Capital Cost

In a general purpose batch production facility, capital equipment and labor are shared among a large number of products, since the requirements for a particular product are not large enough to tie up all the available equipment. As discussed earlier, the flexibility of a job shop is achieved at the cost of reduced levels of equipment utilization. This can be thought of as capacity loss due to sharing. Increased machine utilization resulting from the adoption of robotics can be viewed as recouping some of the capacity that was lost as a result of capital sharing. In many instances, increases in output would cut fixed cost per unit produced drastically, despite the added expenditures for the robots and for the accompanying manufacturing systems. The other major impact of programmable automation on capital cost would be a reduction in the long lead times incurred in the sharing context, which would reduce some part of the inventory cost, as mentioned earlier.

The capital costs of employing flexible manufacturing systems at present, are typically somewhat greater than those of conventional types of automation.[18] Currently, computer controlled, flexible automation is expensive. However, this differential might be reduced to some extent—or even eliminated—if (or when) such systems are built out of relatively standardized elements, such as mass-produced robots and mini- and microcomputers, as opposed to being custom built, as are most specialized production facilities.

---

typically a a sixty-month lead time. Manufacturing engineering typically begins in the period between forty-five and thirty months prior to introduction. In the last thirty months, manufacturing facilities are designed, tested, and built. Simultaneous to plant construction, the product design is refined and further tested. An improvement in machine tool lead times would only affect the last thirty months of the planning cycle. However improving the whole design engineering process could affect lead times as far back as 60 months, or more.

[18] However, the flexible manufacturing systems typically have a greater range of capabilities than the equipment they replace. While overall capital cost might be higher, cost per unit of capability (if this could be measured) might well be lower for the computer-controlled flexible manufacturing systems.

In general, if capacity could be increased significantly, flexible manufacturing systems could reduce both fixed and variable costs of batch production per unit. Figure 6–8 shows how robotics and CAM promise to shift the existing average unit cost curve envelope of Figure 6–4 closer to the ultimate lower limit (materials cost) over much of the spectrum of the production rates, particularly the mid- to high-volume range.

THE CONSEQUENCES OF DRAMATIC IMPROVEMENTS IN PRODUCTIVE CAPABILITIES

Suppose, as we suggest, unit cost in batch and custom manufacturing could be reduced below current levels, mostly as a result of expanding the capacity (and output) of existing facilities.[19] There are several consequences for employment, depending on how demand for the product changes as price declines. This relationship is given by a parameter referred to as a product's ("own") price elasticity of demand.[20] A distinction is usually drawn between two cases:

Inelastic   A 1 percent reduction in price leads to less than a 1 percent increase in quantity demanded. As a result, price cuts decrease total revenue.

elastic   A 1 percent reduction in price leads to more than a 1 percent increase in quantity demanded. Price cuts result in an increase in total revenue.

The price elasticity of demand is unitary if a 1 percent decrease in price leads to a 1 percent increase in quantity demanded. Price cuts leave total revenue unchanged.

---

[19] Much of the savings could, in principle, be achieved without eliminating labor. However, the higher machine utilization rates can be achieved only using by computers and robots to control the flow of work within the whole factory, eliminating the need for much of the "hands-on labor," which in turn eliminates worker-related slowdowns and bottlenecks. If capacity increases are achieved, it would be profitable to pay some of the current workers just to stay out of the way in order that the machines can be more fully utilized. However, this is unlikely to be the most productive, or socially acceptable, use of human resources.

[20] The demand for a product depends on its own price, as well as the price of other products that could be used as substitutes. In the following discussion, we assume that only the price of the product in question is varying and that prices of other products and the values of other important variables, such as income levels, remain constant.

Figure 6–8.   Potential Impact of Robotics and CAM on Cost of Batch Processes.

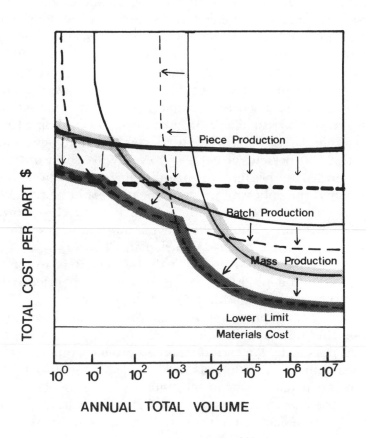

Suppose that plants within a particular industry could achieve dramatic increases in output and consequent decreases in unit cost in robotic plants, but that demand for the product is insensitive to price changes (inelastic). Because revenues also decline, producers are little, if any, better off then before and employment will not rise. This might be the case in certain mature industries, such as steel and rubber and some types of intermediate goods (e.g., auto parts). It might also be true of consumer goods (e.g. T.V's) whose markets are already "saturated." If, despite lower prices, demand remained at or near current levels, fewer plants would be needed to meet current requirements. Less productive, older facilities would most likely be shut down. According to recent data (summer 1982), many industries are currently producing at less than 70 percent of capacity, even with their present technology, primarily because of insufficient orders. In a depressed economy, the only way production costs could be reduced by increasing output is to consolidate facilities and close down the least efficient plants. This would have a more dramatic effect on job displacement than discussed earlier in Chapter 5.

In some cases, though, demand will be sensitive to price changes (elastic). The producer would benefit since price cuts would be more than compensated for by increases in sales, increasing revenues, and profits. What happens to total employment within the industry? This depends on just how sensitive demand is to price cuts. We consider some simple examples to explore whether price cuts could reasonably be expected to lead to net employment increases within an industry using robots.

First, take the case of a relatively modest price cut. Suppose, for the sake of argument, that firms within an industry retrofit robots into existing facilities. They realize a 10 percent decrease in production labor cost and a 25 percent increase in throughput, at the cost of a 5 percent increase in annual capital outlays. Based on these assumptions, labor requirements per unit of output would decrease from 1.0 to 0.72, a 28 percent decrease. Given this decrease in unit labor requirements, demand within this industry would have to increase by $(1.0/.72 - 1.0)$ or by 39 percent to keep the same number of workers employed as previously. To calculate the reductions in unit cost, which we assume are passed on as price cuts, we figure that the cost proportions in this industry are representative of several industries within the fabricated metals sector (SIC 34).[21] Assuming the improvements and costs mentioned above, this industry would realize a 21 percent reduction in unit cost. The price elasticity necessary to induce enough increase in demand to keep all the displaced people employed is given by

$$\frac{39\% \text{ increase in output}}{21 \% \text{ decrease in price per unit}} = -1.86$$

In other words, for each 1 percent *decrease* in price, there would have to be a corresponding 1.86 percent *increase* in quantity demanded to generate enough additional employment so that displaced workers could remain within the industry. For comparison purposes, we note that the elasticity for household appliances has been estimated to be roughly -2.0, which makes it one of the most price responsive of all consumer goods. We would not expect capital goods or most other products of the metalworking industry to be as price responsive as consumer appliances. Thus, even a relatively "modest" cost improvement achieved by installing robots would require that the demand for an industry's products be highly elastic if we were to hope that all displaced workers were to remain within the industry.

Manufactured products are classified as either intermediate goods or final goods, depending on to whom they are eventually sold. Sales of material inputs to manufacturing industries or to other sectors (e.g., construction) are classified as intermediate goods. Sales to any of the four components of final demand—personal consumption, gross private domestic investment, exports, or government purchases (including defense)—are classified as final goods. Investments in capital equipment are credited to final demand as part of gross private domestic investment. The portion of metalworking shipments to intermediate uses and to capital investment are shown in Table 6–21. Almost all primary metal products (both domestically produced and imported) are used as intermediate inputs. So are almost all fabricated metal products,

---

[21] We assume the following cost proportions: production labor cost equals 23 percent of value of shipments, nonproduction labor cost equals 7 percent, capital cost equals 20 percent, and materials cost equals 50 percent. Note these proportions are somewhat different than the average proportions for all metalworking industries (SIC 33-38) given in Figure 6–5)

Table 6–21. Fraction of Metalworking Shipments to Intermediate Uses and Capital Investment.

Industries[a]	Intermediate Inputs % of Shipments	Capital Goods % of Shipments
Primary metal products		
Primary iron and steel	105.7[b]	0.0
Primary nonferrous metals	105.7[b]	0.3
Fabricated metal products		
Metal containers	98.5	0.3
Heating, plumbing, and structural metal products	85.3	7.7
Screw machine products and stampings	90.3	0.0
Other fabricated metal products	85.6	4.5
Machinery, except electrical		
Engines and turbines	50.6	28.4
Farm and garden machinery	17.3	76.7
Construction and mining machinery	19.5	49.4
Material handling machinery and equipment	35.8	57.8
Metalworking machinery and equipment	38.1	52.0
Special industry machinery and equipment	19.2	72.1
General industrial machinery and equipment	64.2	26.9
Miscellaneous machinery and equipment	90.6	0.9
Office, computing, and accounting machines	20.2	53.3
Service industry machinery	56.1	25.7
Electrical and electronics machinery, equipment and supplies		
Electrical industrial equipment and apparatus	56.7	31.5
Household appliances	15.0	14.9
Electric lighting and wiring equipment	75.6	1.5
Radio, television, and communication equipment	21.2	25.1
Electronic components and accessories	83.6	0.2
Miscellaneous electrical equipment and supplies	54.8	10.2
Transportation equipment		
Motor vehicles and equipment	33.3	25.6
Aircraft and parts	22.6	12.0
Other transportation equipment	12.6	46.7
Instruments		
Scientific and controlling instruments	39.5	29.9
Optical, opthalmalic and photographic equipment	31.2	31.9

[a] Industry breakdowns follow the input-output classification scheme, not the SIC scheme.
[b] Includes both domestic production and imports.
Source: Bureau of Commerce. See Ritz (1979: Table 1).

miscellaneous machinery products, and electrical components and accessories. Capital equipment originates mostly but not exclusively from the machinery, except electrical industry. Products sold to the government for national defense are credited to the government purchases component of final demand, which explains why large proportions of aircraft and of communication equipment are neither intermediate nor capital goods. Of all the products of the metalworking sector, only household appliances, motor vehicles, other transportation equipment (bicycles, motorcycles, campers, etc.), and miscellaneous electrical products ship most of their output directly to consumers.

Because most metalworking products are either material or capital inputs for other industries (including metalworking itself), cost savings within the metalworking sector have "ripple effects" on the prices of all goods and services throughout the rest of the economy. (For the same reasons, cost increases in metalworking drive up prices throughout the economy.) It follows that demand for most of these goods, with the exception of autos, appliances, and other goods sold directly to consumers, is "derived" indirectly from the demand for final goods and services. For most metal products, it is unlikely that demand would be highly elastic. Unitary elasticity of demand (the percentage increase in quantity demanded equals the percentage decrease in price) is the most reasonable for most intermediate goods, lacking other data. Given a price elasticity of unity, a 21 percent reduction in price per unit would induce a 21 percent increase in output, with no change in revenues. With the reduced unit labor requirements (as above), this outcome would still leave 13 percent of the workers previously employed in the industry displaced.[22] Suppose, though, that demand for the product *is* highly responsive to price, as might be the case for automobiles, appliances, and other personal consumption goods. If the price elasticity of automobiles were -2, a 21 percent reduction in total cost would induce a 42 percent increase in sales. Even with the reduced unit labor requirements as given above, a 42 percent increase in sales would require a two percent increase in employment requirements, and produce additional revenue for the manufacturer as well.

---

[22] .72 labor units/unit of output * 1.21 units of output = .87 units of labor = 13 percent decrease from the base case.

Robot use could easily result in more dramatic reductions, both in unit labor requirements and in prices.[23] Suppose all producers within an industry were to completely refurbish their existing facilities, or even build new plants. For the sake of discussion, assume the new facilities realized a 20 percent reduction in production labor cost and achieved a 100 percent increase in output, at the expense of a 20 percent increase in annual capital outlays. In this case, unit production cost would drop by 50 percent. With the new unit labor requirements, output would have to expand by 250 percent to create enough work to keep all of the displaced workers employed in the same industry, which would require a very high price elasticity (-5). If demand only increased by as much as prices decreased (50 percent), then employment of production workers would drop 40 percent from previous levels.

The above examples suggest that if robotic technologies were to be widely used, not all displaced workers could be expected to be reemployed in their current industries as a result of price reductions and increased demand. The effects of price reductions on demand, and the net employment effect, balancing job displacement and job creation, will vary considerably among the various metalworking industries, depending on the nature of the product and its market. The logical conclusion is that employment of production workers in most manufacturing industries would decrease, despite substantial improvements in productivity within these industries and possible increases in production. This does not mean, however, that total employment in the economy as a whole would decrease. Substantial reductions in the prices of intermediate and capital goods (for example 20 percent in the first example and 50 percent in the second one) should reduce the cost of manufacturing consumer goods and of creating new goods and services, both of which will increase the consumers' real buying power and effective demand for other goods and services. This should create new employment opportunities.

---

[23] As an aside, if demand were to increase by several hundred percent (and in some cases, it might), most established organizations are not prepared to cope with the increased complexity of organizing their business. It would strain the organizational structure, especially information processing capabilities. As a result, producers sometimes purposefully restrain their technological capabilities in order to keep their business "manageable."

There is one important point that has been overlooked in the above examples, and in economic analysis in general. We implicitly assumed that the only way to utilize the "extra" capacity made available by using robotic systems was to increase the output of the goods that are already produced in that factory (or industry). However, there is an option of making greater use of the expanded capabilities and of the flexibility of robotic production systems to produce a wider range of products and to manufacture new, high-performance products. Thus, simply looking at the price elasticity of demand for current products might substantially underestimate the extent to which additional capacity could be utilized. If the benefits of robotics and of other types of programmable manufacturing technologies are to be fully exploited, we strongly believe there needs to be a concurrent emphasis on the development of new products to utilize the expanded capabilities and the greater capacity. A new strategy that places much more emphasis on product performance and less on standardization and cost reduction might require an abrupt shift in many existing corporate strategies. The need for such a strategy change and its implications for our future economic growth and survival are discussed in more detail in Chapter 7.

*Will the benefits of robotic technology be fully appreciated and exploited by today's manufacturing management?*Almost all of the existing installations of robotics and of flexible manufacturing systems have been motivated by the desire to reduce the cost, principally the labor cost, of producing exisiting goods. To date, robotic production technologies have had little effect on product design and development and on marketing strategy. This might suggest that manufacturers do not share our view that the major benefits of robotics can only be realized if there is a concurrent emphasis on product as well as process innovation. If this were true, the implications are that the amount of job creation might be small in comparison to displacements. But past trends with respect to the motivations for and uses of robotics might be a misleading indicator of future applications. To date, producers have had relatively little experience with programmable manufactúring technologies, and it is to be expected that the initial applications are motivated by some of the more conservative and easily realized goals. However, there are already strong indications that designers and strategists within some of the major manufacturing companies are giving serious attention to integrating developments in robotics and other manufacturing technologies with product development and overall corporate strategy. If this development comes to be standard industry practice, as opposed to the exception it appears to be now, then there would be reason to expect that the widespread use of robotic technologies would directly and indirectly create as many, or even more, jobs than it displaces.

To analyze the input of productivity improvements within the metalworking sector on economy-wide growth and inflationary pressures, we characterize the primary economic benefit of robotics as a reduction in the real cost of manufacturing products made in small to medium batches. Capital goods—machine tools and the other types of durable equipment—as well as the parts used within them, are largely batch produced.[24] If the real cost of metalworking products, particularly capital equipment, were reduced, the price of capital goods in relation to final products could be expected to decline significantly over the next half century. This would cause secondary ripple effects on the prices of other manufactured goods and services throughout the economy. Reductions in the real price of producer's durable equipment would reduce real capital cost per unit in the sectors using this equipment. This, in turn, would reduce the real price of final output of mass-produced consumer goods, as well as the real price of output of the nonmanufacturing sectors. Final demand would also be stimulated to some extent, depending on the sensitivity of final demand to price. (For consumer goods, high price elasticities tend to be more the rule than the exception). Lower real costs would also have a beneficial impact on the rate of inflation. Insofar as inflation is caused by "too much money chasing too few goods," an increase in productivity is perhaps the best way to break out of the vicious cycle. Ultimately, such changes would effect other important macroeconomic variables, including the overall level and composition of employment and the level and distribution of income. These second order effects, while less immediate, may have greater ultimate importance than the immediate improve-

---

[24] The major exception is automobiles. Large appliances, such as refrigerators, air-conditioners, and washing machines, are also mass-produced durable goods, but since they are sold to consumers, they are not classified as producers' durable equipment.

ments in labor productivity in manufacturing. It is beyond our present scope to attempt to forecast the detailed nature, the magnitude, or the time phasing of these broader economywide economic impacts.[25]

We expect substantial reductions in the price of intermediate and capital goods to play an important role in facilitating the development of new growth sectors in the economy, including hazardous waste management, biotechnology, undersea and space exploration, and undersea mineral exploration. These sectors would also provide employment. It is important to know if the levels of economic growth required to absorb workers displaced by robotics and other forms of technological changes can be achieved in the economy *as it is now structured*. If the required levels of economic growth (and employment) cannot be achieved as a result of cost saving process improvements in manufacturing, resources may have to be reallocated to encourage the creation of new products or services or the development of new frontiers such as the oceans and space. This would require a reevaluation of traditional policies of stimulating economic growth by encouraging aggregate consumption or aggregate investment.

Will workers be better off as a result of productivity improvements in manufacturing? Nobel laureate Herbert Simon writes (1977: 159), "The main long run effect of increasing productivity is to increase real wages—a conclusion that is historically true and analytically demonstrable." He argues that as long as the rate of interest does not rise, any technological improvement, regardless of whether it conserves capital or labor, must, in the long run, raise real wages. He also concludes that any level of technology and productivity is compatible with any level of employment, including full employment. He discusses these assertions in detail in his book Simon 1977). The key point is that unemployment is a distributional problem.

In 1980, the Organisation for Economic Co-operation and Development (OECD), completed a major effort to re-evaluate the links between research, technology and economic growth in the

---

[25] There is an extensive literature devoted to pursuing such questions which has been developed by economists interested in measuring technological change, and by researchers interested in the impacts of energy price increases on economic growth. For example, see Sato and Ramachandran (1980), and Hoffman and Jorgenson (1977).

Western industrialized countries (OECD 1980). In looking at the impact of technical advance on the economies of their member countries, the authors comment (1980: 61), "The most striking manifestation of the technical advance has been the almost tenfold increase in measured per capita incomes achieved over the past two centuries in the OECD area." They basically share Simon's view that technological advance and employment growth are compatible with one another. In fact, they suggest that increasing the rate of technical advance would reduce unemployment. The OECD report explains (1980: 79-80):

> Many people tend to think that it is harder to sustain full employment when technical advance is rapid than when it is slow, simply because output must grow faster in the former case to keep up the productivity growth and the growth of the work force. Historical performance, as well as theoretical analysis, suggest that, on the contrary, it may be easier to maintain full employment when technical advance is rapid than when it is slow, provided the direction of technical change is not adverse. Certainly during the 1950s and 1960s, rapid productivity growth and low unemployment rates tended to go together.
>
> We do not believe that the decline in technical advance and productivity growth associated with the slow-down of the last half dozen years has made the unemployment problem *more* intractable.

The committee of international scholars that prepared the OECD report share our views on the importance of continuing innovation. They observe that absent of the creation of significant new products and production techniques, routine process technical advance does tend to involve a considerable substitution of capital for labor. The implication is that innovation is as central to economic growth as reducing cost. We have already made this point several times and will discuss it further in Chapter 7.

An optimistic assessment of the *potential* payoffs that could be realized from accelerating the use of robotics is given in a recent Joint Economic Committee staff paper by Vedder (1982: 18):

> If things go as expected, everyone should gain through the adaption of the new innovation, a robot: workers in the form of higher wages, consumers in the form of lower prices, and the producer in the form of higher profits.

We do not disagree, but we note that the nature of the potential impacts and benefits are more complex than this statement might suggest.

CONCLUDING COMMENTS

The technology of CAM and robots exists today in scattered locations throughout the industrialized world. How long it will take to go from where we are to the millennial "factory of the future" is not clear. One important point is the fact that the changeover will involve the replacement of hundreds of thousands of obsolete manually operated machine tools by computer controlled versions. The vast market for new high technology processing and materials handling equipment needed with sectors SIC 33-38 will have to be met by firms in the same sectors. This means demand for computer controlled machines and robots will be exceptionally high at least for a decade or so during the transition period and the required higher levels of capital goods output can be met only by initially introducing more productive processes and equipment within the plants of the capital goods producers themselves. (The alternative would be to rely on imports of computer controlled machines and robots from Japan—a strategy that would leave the U.S. machine tool industry permanently crippled.) Later, the benefits of additional productivity will spread throughout the engineering industries and ultimately through the rest of the manufacturing sectors.

Assuming (optimistically, perhaps) that the coming revolution in manufacturing technology will begin in the capital goods sector itself, the combination of CAD/CAM and robotics will make the next generation of capital goods much more productive than the old equipment it replaces. In real terms, the price of capital goods (in relation to the price of output) can be expected to decline fairly sharply in the next half century. Other sectors, such as construction, transportation, and utilities, will also be beneficiaries of robotics via lower cost equipment. There will be a ripple effect on prices of other manufactured consumer goods—in relation to labor and energy input—and final demand will be correspondingly stimulated. These long-term effects, while less immediate, will have greater ultimate importance than the expected (minor) improvements in labor productivity noted earlier.

In addition, robotics and CAM will reduce the benefits of mass production relative to batch production and thereby reduce the existing incentives toward product standardization and against diversity and change. This should have an accelerating effect on innovation during the 1980s and induce a much greater emphasis on product performance.

Discussions on how to improve manufacturing productivity and spur economic growth focus, for the most part, on the quantity of capital available and on the efficiency with which it is used. While better capital utilization is necessary for growth, it is not sufficient to make growth occur. The concept of economic efficiency must be expanded to encompass the notions of adaptability over time, diversity, and rapid change, in addition to the narrowly defined concept of minimizing cost. One of the consequences of achieving *large* increases in productivity will be a need to find new ways of utilizing the extra capacity and of creating employment. We believe this requires a sharply higher rate of innovation for industry.

In summary, we are convinced that robotic technologies hold enormous potential for stimulating economic growth and for securing fulfilling employment for people. Whether or not these technologies could *potentially* make our society better off is not really the right issue to debate. Unquestionably it could, and the fact that technical change in general has increased real wealth and employment is "historically true and analytically demonstrable,' as Professor Simon reminds us. The critical issues, in our minds, are where to employ robotic technologies, how to manage their use, and how to make these changes in ways that will be jointly supported by all levels of management and labor. In short, how not to blow it. A wide range of social problems may erupt if these technologies are insensitively introduced or poorly managed. Because of their complexity and power, learning how to manage robotic technologies will require a considerable effort. In addition, we must learn how to make the coming transition in ways that will minimize social disruption, increase job safety and work satisfaction, and enhance overall quality of life for workers. This leads us to the discussion of policy issues in the next chapter.

REFERENCES

American Machinist. The 12th American Machinist Inventory of Metalworking Equipment, 1976–78. *American Machinist* 122(12), December 1978.

Ayres, Robert U. *Resources, Environment and Economics.* John Wiley and Sons, Inc., New York, 1978.

Ayres, Robert U., and Steven M. Miller. Robotics, CAM and Industrial Productivity. *National Productivity Review* 1(1), Winter 1982.

Borzcik, Paul S. Flexible Manufacturing Systems. In Arthur R. Thompson, Working Group Chairman (editor), *Machine Tool Systems Management and Utilization,* pages 62–74. Lawrence Livermore National Laboratory, October 1980. Volumn 2 of the Machine Tool Task Force report on the Technology of Machine Tools.

Bureau of Labor Statistics, U.S. Department of Labor. *Handbook of Labor Statistics.* Government Printing Office, Washington, D.C., 1980.

Bureau of Labor Statistics, U.S. Department of Labor. *Occupational Employment in Manufacturing Industries, 1977.* Government Printing Office, Washington, D.C., 1980. Bulletin 2057.

Bureau of the Census. *Survey of Plant Capacity, 1979.* Government Printing Office, Washington, D.C., 1980. Current Industry Report Series, Report MQ–C1(79)–1.

Bureau of the Census, Industry Division, U.S. Department of Commerce. *1977 Census of Manufacturers.* Volume MC77–I–Part 1 and 2: *Industry Statistics.* Government Printing Office, Washington, D.C. 1981.

Bylinsky, Gene. A New Industrial Revolution Is Under Way. *Fortune,* 5 October 1981.

Capdevielle, Patricia, and Donato Alvarez. International Comparisons of Trends in Productivity and Labor Costs. *Monthly Labor Review*:14–20, December 1981.

Carter, C.F., Jr. Trends in Machine Tool Development and Application. In *Proceedings of the Second International Conference on Product Development and Manufacturing Technology,* pages 125–141. Macdonald, London, 1972.

Cook, Nathan H. Computer-Managed Parts Manufacture. *Scientific American* 232(2):22–29, 1977.

Dallas, Daniel B. Major Causes of Productivity Loss. *Manufacturing Engineering* 87(3):79–82, September 1981.

Drucker, Peter. Japan Gets Ready for Tougher Times. *Fortune* 102(9):108–14, 3 November 1980.

Eversheim, Walter. Economic and Technical Limits of Cutting Technologies. In Arthur R. Thompson, Working Group Chairman (editor), *Machine Tool Systems Management and Utilization,* pages 8.16–1 to

8.16–14. Lawrence Livermore National Laboratory, October 1980. Volumn 2 of the Machine Tool Task Force report on the Technology of Machine Tools.

Fraumeni, Barbara M. and Dale W. Jorgenson. The Role of Capital in U.S. Economic Growth. In George von Furstenberg (editor), *Capital, Efficiency and Growth*. Ballinger Publishing, Cambridge, Mass., 1980.

Fromm, Gary. Research on Capital and Productivity. In John D. Hogan and Anna M. Craig (editors), *Dimensions of Productivity Research, Vol. 1,* pages 109–114. American Productivity Center, Houston, Texas, 1980. Proceedings of the Conference on Productivity Research held at the American Productivity Center, Houston, Texas, April 21–24, 1980.

Hoffman, Kenneth C., and Dale W. Jorgenson. Economic and Technological Models for Evaluation of Energy Policy. *Bell Journal of Economics* 8(2):444–466, Autumn 1977.

Kalmbach, P., R. Kasiske, F. Manske, O. Micker, W. Pelull, and W. Wobbe-Ohlenburg. Robots Effect on Production, Work and Employment. *Industrial Robot* 9(1):43–46, March 1982.

Kendrick, John W. Productivity Trends in the United States. In Sholmo Maital and Noah M. Meltz (editors), *Lagging Productivity Growth: Causes and Remedies,* pages 9–30. Ballinger Publishing, Cambridge, Mass., 1980.

Kendrick, John W., and Elliot Grossman. *Productivity in the United States: Trends and Cycles*. Johns Hopkins Press, Baltimore, Md., 1980.

Markowitz, Harry M., and Alan J. Rowe. The Metalworking Industries. In Manne, Alan S., and Harry M. Markowitz (editors), *Studies in Process Analysis*. John Wiley and Sons, New York, 1963.

Mayer, John E., and David Lee. Estimated Requirements for Machine Tools During the 1980–1990 Period. In Arthur R. Thompson, Working Group Chairman (editor), *Machine Tool Systems Management and Utilization,* pages 31–41. Lawrence Livermore National Laboratory, October 1980. Volumn 2 of the Machine Tool Task Force report on the Technology of Machine Tools.

Mayer, John E., and David Lee. Future Machine Tool Requirements for Achieving Increased Productivity. In Arthur R. Thompson, Working Group Chairman (editor), *Machine Tool Systems Management and Utilization,* pages 8.4–1–8.4.12. Lawrence Livermore National Laboratory, October 1980. Volumn 2 of the Machine Tool Task Force report on the Technology of Machine Tools.

Merchant, M. Eugene. The Inexorable Push for Automated Production. *Production Engineering*:44–49, January 1977.

1981–82 Economic Handbook of the Machine Tool Industry. National Machine Tool Builders Association, McLean, Va., 1981.

Norsworth, J.R. *Recent Productivity Trends in the U.S. and Japan*. Technical Report, Office of Productivity and Technology, Bureau of Labor Statistics, Washington, D.C. April 2, 1982. Testimony prepared for the United States Senate Subcommittee on Employment and Productivity of the Committee on Human Resources, 2 April 1982.

Norsworthy, J.R., and Michael Harper. The Role of Capital Formation in the Recent Slowdown in Productivity Growth. In Dogramaci, Ali, and Nabil Adam (editors), *Aggregate and Industry Level Productivity Analysis*, pages 122–148. Martinus Nijhoff Publishing, Boston, Mass., 1981.

Norsworthy, J.R., M.J. Harper, and K. Kunze. *The Slowdown in Productivity Growth: An Analysis of Some Contributing Factors. Brookings Paper 2*, Brookings Institute, Washington, D.C., 1979.

National Research Council. *Measurement and Interpretation of Productivity*. National Academy of Sciences, Washington, D.C., 1979. Report of the Rees panel to review productivity statistics.

OECD. *Technical Change and Economic Policy*. Organisation for Economic Co-Operation and Development, Paris, France, 1980.

Office of Technology Assessment. *U.S. Industrial Competitiveness: A Comparison of Steel, Electronic and Automobiles*. Technical Report OTA–ISC–135, Office of Technology Assessment, July 1981.

Real, Bernard. *The Machine Tool Industry*. Technical Report, Organization for Economic Cooperation and Development, August 1980. Sector report for the OECD study on Technical Change and Economic Policy.

Ritz, Philip M. The Input-Output Structure of the U.S. Economy, 1972. Survey of Current Business:34–71, February 1979.

Sadler, George E. *Productivity Prospectives, 1981 edition*. Technical Report, American Productivity Center, Houston, Texas 1981.

Sato, Ryuzo, and Rama Ramachandran. Measuring the Impact of Technical Progress on the Demand for Intermediate Goods: A Survey. *Journal of Economic Literature* XVIII:1003–1024, September 1980.

Simon, Herbert A. *The New Science of Management Decision*. Prentice-Hall, Inc., Englewood Cliffs, 1977. Revised edition.

Stone, Richard. *Input-Output and National Accounts*. Organisation for European Economic Cooperation, Brussels, Belgium, 1962.

Tesar, Delbert. Mission-Oriented Research for Light Machinery. *Science* 201:880–887, 8 September 1978.

Tesar, Delbert. Mechanical Technology R&D: The Tragic Need. *Mechanical Engineering* 102(2):34–41, February 1980.

Thompson, Arthur R. Introduction and Summary. In Arthur R. Thompson, Working Group Chairman (editor), *Machine Tool Systems Management and Utilization,* pages 1–26. Lawrence Livermore National Laboratory, October 1980. Volumn 2 of the Machine Tool Task Force report on the Technology of Machine Tools.

Usher, Dan (editor). *National Bureau of Economic Research Studies in Income and Wealth.* Volume 45: *The Measurement of Capital.* University of Chicago Press, Chicago, Ill., 1980.

Vedder, Richard K. *Robotics and the Economy.* Technical Report, Joint Economic Committee, Congress of the United States, March 1982.

Vietorisz, Thomas. *UNIDO Monographs on Industrial Development.* Volume 4: *Engineering Industry.* United Nations Industrial Development Agency, Vienna, Austria, 1969.

Furstenberg, George M. (editor). *Capital Investment and Saving.* Volume III: *Capital, Efficiency, and Growth.* Ballinger Publishing Company, 1980. Sponsored by the American Council of Life Insurance.

Westphal, Larry E., and Yung W. Rhee. The Methodology of Investment Analysis for a Non-Process Industry. In Stoutjesdijk, Ardy, and Larry E. Westphal (editors), *Industrial Investment Analysis under Increasing Returns.* The World Bank, Washington, D.C., 1983. Forthcoming.

7 POLICY IMPLICATIONS

At first sight, it appears that there are two basic areas of policy that require urgent attention by national, state, and local government and by industry. One is relatively narrow; it is concerned in part with compensating for inequities in the allocation of social costs resulting from the coming introduction of robots and computer-assisted manufacturing (CAM), especially in the metalworking industries. Many of the responses to this problem are also applicable to the related problem of developing human knowledge and skills that will be most needed in the postindustrial society into which our society is evolving. Both of these concerns can be included under the heading of *human resources policy*.

A broader arena for policy analysis relates to the problem of productivity at the sectoral and national levels. The urgency of this problem arises from the fact that the United States' economy is no longer self-sufficient and can no longer ignore the rest of the world. As a nation, we must now compete with other nations in areas ranging from geopolitical to economic. Military power, of course, rests on economic foundations. Automatic U.S. economic dominance can no longer be taken for granted.

Partly because the United States has been highly industrialized for a long time, and partly because industrialization originally occurred in an era of plentiful natural resources, the U.S. has rather suddenly, over a period of three decades, become a massive importer of many industrial raw materials, including petroleum, iron ore, copper, manganese, chromium, cobalt, and platinum group metals. (In contrast, the U.S.S.R. is largely self-sufficient.) To pay for these imports the U.S. economy must successfully export manufactured goods—in competition with Europe, Japan, and a number of other countries. But the U.S. industrial establishment is having increasing difficulty in the trade competition, not only in foreign markets but even at home.

In particular, the industries producing basic metals (and chemicals) and standardized products such as automobiles, tires, consumer electronics, cameras, watches, and so on are experiencing increasingly severe difficulties. The reason is fairly easy to diagnose; in these sectors competition in the marketplace is largely on the basis of price, and the lowest price requires the lowest cost of mass production.

Given that the technology of production in these industries is itself largely standardized, the lowest production cost is likely to be found in a country with (1) very cheap raw materials, (2) cheap (perhaps subsidized) capital, and (3) cheap labor. In Japan raw materials have not been cheap but capital has been relatively inexpensive (at least, compared to the United States) both in terms of direct man-hour cost and in terms of labor quality motivation, skill, and effective deployment. Japanese unions, for instance, do not demand restrictive "work rules" that create phantom jobs.

For reasons that were briefly touched on in Chapter 6, the loss of mass production industries to overseas locations is unlikely to be reversed in the United States, even by substantial subsidies to capital and upgrading of plant and equipment. At best, the migration of basic industry overseas can be slowed down somewhat. The broad implications are clear: the United States must find other kinds of manufactured products to export, or it must sell more services abroad or cut down very sharply on imports, or all three.

What manufactured products might the United States continue to produce competitively for export? Production must be in the "batch" mode because the technology *cannot* be standardized. This, in turn, implies a continuing rapid rate of technological innovation in both the product and its production techniques. Otherwise standardization would be inevitable and rapid. Military aircraft and computers are examples of such products.

For reasons explained in Chapters 2 and 3, robotics and CAM are particularly well suited to the batch mode of production, especially when product design is constantly evolving. Thus, robotics and CAM will contribute significantly to the ability of the United States to shift its industrial system away from concentration on low-cost mass production of standardized products to batch production of rapidly evolving products. To accomplish this, a number of fundamental institutional changes must occur. These include:

- *Long-term strategic planning.* Just as corporations plan for the future, so must the nation. Success will require concentration of resources (including federal R&D funds) in promising newer areas and encouraging older mature industries to do likewise. Here the major problem is to resist political pressure for protectionism and federal subsidies (or ownership) "to save jobs." A more effective human resources policy would help to reduce these pressures.
- *Avoidance of the "imperial" syndrome.* It is tempting to think that the United States must, at all costs, protect its "lifelines" to guarantee access to resource exporters such as Zaire (cobalt), South Africa (chrome, manganese, platinum), Saudi Arabia, and Iran (petroleum). This doctrine becomes a justification for maintaining an enormous military establishment and subsidizing many unpopular regimes. It can quickly backfire politically as we discovered in Iran. In addition, continued dependence on exhaustible resources delays the development of technological substitutes.
- *Fostering innovation.* Both the generation and the adoption of new technology must be encouraged. The major obstacles are in the latter area: both passive and active forms of resistance to change must be overcome. The key is incentives. In industry, more emphasis must be put on rewarding workers and managers for long-term results. Most important, workers (both salaried and hourly) must be relieved of fears of technological displacement and unemployment. A new *workers' bill of rights*, focused on mid-career retraining and relocation assistance, should replace the present unemployment insurance system with its perverse incentives.

Evidently the human resources issues are in fact central to the productivity problem—indeed, the problem of international competition as well. We focus primarily on human resources and technology policy, commenting only briefly on other aspects of U.S. economic policy. We also discuss the role of R&D compensation and promotion policy in the private sector.

HUMAN RESOURCES POLICY

It is unfortunate that both industrialists and workers tend to think of robots almost exclusively as "labor-saving" machines. That industry does so is revealed clearly by the fact that, despite many public assurances that "no worker has lost his job to a robot,‘ the financial justification for purchasing a robot nowadays is invariably (and often exclusively) based on *labor hours saved*. Workers and their unions, not being blind or stupid, get to know this. Blue-collar operatives, in particular, are beginning to feel a chill apprehension about their future. Understandably, workers in many industries are beginning to seek means of resisting the expected onslaught of "steel-collar workers." The mechanisms being considered include tighter work rules, notification requirements, increased labor voice in shopfloor decisions, inclusion of robot operators and programmers in the bargaining unit, shorter work weeks, "personal paid holiday," and so on.

Unfortunately, all these mechanisms will, to the extent they are adopted, further increase labor costs, further reduce productivity, and exacerbate the long-run problem of U.S. noncompetitiveness. Worse, none of these measures will really be effective in protecting jobs or in reducing workers' fears. In fact, many of the restrictive mechanisms noted above are already part of United Auto Workers (UAW) contracts with the major automobile manufacturers. Indeed, the high costs of the "protections" in the UAW contract are partly responsible for the permanent closing of many plants and the loss of all jobs in these plants—regardless of length of service to the firm. Even priority for job openings in other plants is small compensation to long-service employees, especially when seniority is not transferable to other bargaining units.

As a rule, there are severe restrictions on the transferability of seniority rights for promotion and for protection against job layoffs.[1] Seniority rights in these two critical areas are usually forfeited if the worker transfers out of the bargaining unit.[2] In some contracts, seniority is specific to particular work areas within the plant. There are even cases where these rights are retained only if the worker remains within a specific occupation within the bargaining unit. Locals of the same union in different plants will not allow a worker to transfer seniority rights for promotion and for protection against layoffs. Needless to say, seniority is never retained if the worker joins another union.

Nontransferability of seniority is one of the most effective impediments to labor mobility, since it inhibits upgrading of skills, especially among older employees. Yet the problem we face as a nation is, precisely, to convert several million workers whose existing skills are obsolescent to other, higher skills that will be in demand.

There are other serious difficulties to be overcome. The existing unemployment insurance (UI) system, the existing *safety net* for dislocated workers, is one. This system is expected to expend $23 billion in fiscal year 1982. Eligibility requirements vary across states, but as a rule benefit payments are determined primarily by the worker's earnings within the last four to six calendar quarters.[3] A lifetime's worth of work experience accumulated before the year to year and a half "qualification period" is not considered in determining UI benefits. The established worker with many years on the job, children, and a mortgage is usually left high and dry if his plant closes—as has happened many times in the past year. In our view, a program that provides approximately the same unemployment benefits to a worker with one year's experience as to a worker with 30 years on the job is somewhat perverse. The program also encourages abuse. After several months of working, a drifter can quit and live "free" on benefits

---

[1] Seniority rights for other privilages, such as vacation preferences, health care, pensions, or for overtime, are more easily transferred across bargaining units within the same company.

[2] If a union such as the UAW has national agreements with a large company, there are exceptional circumstances under which seniority rights can be transferred. In some instances when this has happened in the past, the transplanted worker was greeted with hostility by other workers in the plant to which he transferred.

[3] The standard period of benefits is 26 weeks. If a state has a higher than average rate of unemployment, benefits may be extended by an additional 13 or 26 weeks.

for the next six months, and repeat this cycle indefinately. Such a program gives no incentive to people with obsolete or unstable jobs to learn other skills.

The recent federal training program (CETA), which will be phased out by the end of 1982, has been another obstacle. Its primary emphasis was, and is, on providing "entry-level" skills (and work habits) to disadvantaged minorities (Table 7-1). For the most part, it was not designed to assist experienced semiskilled workers to upgrade their skills or develop new ones. Indeed, some of the existing training programs run by CETA tend to concentrate on training the same low-level skills that are most likely to be displaced by robots and other forms of automation. Federally funded vocational training programs also train new job entrants

Table 7-1 CETA-Sponsored Training And Retraining Programs.

Program Characteristics	HIRE Title III	OJT	JC Title IV	STIP Title II
Services	On-the-job training	On-the-job training	On-the job training	On-the-job training
	Classroom Intitutional training	Classroom	Classroom Vocational education	Classroom Institutional training
Funding	DOL	DOL	DOL	DOL
Clientele	Veterans	Economically disadvantaged	Disadvantaged youth	Skilled workers seeking new skills
Occupations	Metalworkers	Welders	Machine operators	Welders
	Tool and die makers	Assemblers		Assemblers
	Machine operators	Other operatives	Metalworkers	

Key
DOL: Department of Labor.
HIRE: Help through Industry Retraining and Employment.
OJT: On-the-Job Training.
JC: Job Corps.
STIP: Skills Training Improvement Program.

   Source: Carnegie-Mellon (1981: 128).

in some of the same areas that have been identified as prime candidates for robots. The six metalworking occupations shown in
Table 7–2 accounted for just under 3 percent of all vocational education enrollments in fiscal year 1978.

What needs to be done? At the federal government level, an
"active manpower policy" is needed to replace the existing UI
and CETA programs. In particular, the role of UI as a "wage
loss replacement" program must be reconsidered. What is
needed, in our view, is a system that provides transitional benefits for dislocated workers in proportion to length of service and
that provides strong incentives and financing for retraining and
education. Some of the essential components are already part of
experimental government programs currently operating on a
small scale. For example, the Trade Adjustment Assistance
(TAA) program, in operation since 1974, provides income support and training benefits for workers who have lost their job as
a result of import competition. For the most part, the program
serves manufacturing workers, many of whom are 45 and older.
Like the UI program, the income compensation given to eligible
workers is independent of length of service beyond the qualifica-
tion period. In contrast to the UI program, income benefits are
extended if the worker is enrolled in any type of job related vocational training. In addition, the TAA program pays tuition for
training programs, as well as as job search and a job relocation
allowance.

Table 7–2 Enrollments and Completions in Public Vocational
Education in Selected Metalworking Occupations: National
Totals: Fiscal Year 1978.

Occupations	Enrollments	Completions
Machine shop occupations	117,069	32,588
Machine tool operations	14,232	3,347
Sheet metal	45,694	6,571
Welding/cutting	205,486	51,722
Tool/die making	8,475	2,369
Other metalworking occupations	58,709	17,548
Totals	449,665	114,285

Source: Bureau of Labor Statistics (1980).

Several options within the UI system have been proposed over the past decade or so.

- Allow the UI system to loan money to laid-off workers for retraining, the loan to be repaid by a surcharge on the individual's UI contribution after reemployment.
- Tax-free employer and employee contributions to a fund to be used for retraining (or as a pension supplement after the worker retires).
- UI taxes could be abated for employers who outplace or retrain workers as an alternative to layoff.

All these schemes have some merit, but none has yet been subjected to thorough analysis with regard to eligibility, solvency, efficiency, or "the new federalism."

Options outside the UI system include:

- The proposed Job Training Act (to replace CETA) could be modified to include an explicit focus on retraining of displaced workers.
- Federal support (through the Department of Education) for adult vocational educational programs could be increased significantly.

The complex structure of the federal government, both congressional committees and executive agencies, tends to favor piecemeal "tinkering" with existing programs rather than new initiatives. Nevertheless, the economic stakes are such that a single comprehensive program might well be the best answer. Such a program might be called the Workers' Bill of Rights (recalling the GI Bill of Rights introduced after World War II). The key features of such a plan would include the following:

- All employed workers and their employers would contribute (tax free) to a fund (replacing existing UI) to finance education and training. Each worker would build equity in the system in proportion to the contribution made in his behalf. Credit would be transferable from job to job. The fund would pay tuition and fees at recognized vocational schools and colleges and provide supplementary income in times of job transition.

- Accrued benefit rights would be proportional to cumulative contributions. Workers who never exercise their retraining rights would be entitled to convert them into a pension supplement annuity at retirement.

- Short-service workers with little equity who are displaced involuntarily would be entitled to borrow from the fund with repayment deducted from later contributions.

- Ideally, the program should be self-financing. However, an initial "endowment" from general revenues should not be ruled out, notwithstanding current budgetary stringencies.

- The problem of finding jobs for handicapped, disabled, unskilled, uneducated, and disadvantaged minorities who have never been productively employed should be dealt with by different means. One possibility is a new version of the Civilian Conservation Corps (CCC) of the 1930s. Another is tax subsidies for firms willing to hire and train inexperienced workers. Reagan's "Enterprise Zone" concept may also have some promise. Existing welfare laws with their perverse disincentives to work in low-paying or part-time jobs should be replaced by a negative income tax.

The Workers' Bill of Rights would emphasize mid-career retraining and educational upgrading. Two or three decades from now it should not be unusual for men and women in their thirties, forties, and fifties—with families—to be attending vocational schools, colleges, and graduate schools full-time for 1-year, 2-year, and even 3-year programs to earn formal degrees. One or more major career changes during a working life should be the rule rather than the exception.

The likely social consequences of major losses of industrial jobs—whether due to economic depression, migration of manufacturing to oversea export platforms, or to robotization—are obviously very severe. A foretaste of these consequences has already been experienced in the older northeastern industrial heartland of the United States. The worst scenario is a disappearance of the well-paid blue collar middle class, resulting in a bifurcated dual society consisting of low wage, unskilled service jobs on the one hand, and high wage, "elite" professional jobs on the other. We think this kind of division would sharply decrease social mobility in our society and would have ominous long-range implications for our democracy. The only way to steer away from such a grim prospect, in our view, is to undertake a massive national investment in both education and skill training across the whole spectrum of ages and social backgrounds—an updated version of the GI Bill of Rights for our workforce.

NATIONAL R&D POLICY

Traditionally, national technology policy is concerned with the allocations of federal support for research and development. Much of the debate, year in and year out, concerns the role of basic versus applied research. A second running debate is concerned with the proper role of federal government vis à vis the private sector in supporting research directed at economic objectives. Conservatives tend to argue that the government "crowds out" the private sector, while liberals believe that the government can play a useful leadership role. There can be little doubt that experience in the United States, especially in the energy area, suggests that the conventional system of piecemeal research contracting by a federal agency is not as effective as it might be. On the other hand, federal cost sharing, with rights to exploit the results remaining with the contractor, may work quite well. Research consortia of the sort frequently organized in Japan by the Ministry of International Trade and Industry (MITI) can also be organized in the United States (as the semi-conductor and computer manufacturers have recently demonstrated).

We specifically suggest that such a consortium be created for the explicit purpose of designing and building a robotic automobile engine plant, integrated with computer-aided design (CAD). The objective of the effort should be to create a new flexible batch production technology for engines permitting engine designs to be altered and improved on a continuing basis. Once having been built and demonstrated, such a plant could be put up for sale to one of the existing U.S. auto firms, or it could become an independent engine supplier to the industry.

The cost of such a project would be significant—perhaps several billion dollars—but success in this venture could be critical to the long-term viability of the U.S. auto industry. It would, on the other hand, be a less expensive and more economically significant project than, for instance, the space shuttle.

CORPORATE R&D POLICY

In general, corporate goals include survival, growth, and profitability. Specific quantitative targets are typically chosen to guide internal resource allocation, to identify and reconcile conflicts, to monitor performance on a continuous basis, and to give all employees a sense of purpose. The function of R&D in a modern corporation is to provide a flow of technology in support of these goals. Long-term survival and growth requires the introduction of new products to replace obsolete ones.[4] Profitability, on the other hand, may be improved either by increasing demand (where there is excess capacity) or by reducing costs. R&D contributes to the first by product improvement and to the second by process improvement.

Clearly, technology may be acquired from two basic sources: internal (i.e., from within the R&D labs) or external. In the past, most major U.S. firms have attempted to develop almost all their own technology internally. In a few cases, this strategy was perhaps justified on the grounds that the United States in general, and the firm in particular, already led the world and, therefore, had little to gain from the outside. Such a claim can no longer plausibly be made by many U.S. firms, if by any. (The one possible exception might be Bell Laboratories.)

It makes a limited amount of sense for a firm with a technological advantage over all its competitors (domestic and foreign) to screen out most offerings from the outside world on the grounds that (1) no firm can take up all options or investigate everything and (2) the firm is already exploring the most promising avenues. On the other hand, it does *not* make sense for a firm that is technologically "one of the pack" to behave in this way. Nevertheless, traditional corporate R&D departments continue to raise effective barriers against accepting ideas from outside, rather than

---

[4] The product life cycles may be measured in months (e.g., some ladies fashions) or in centuries (e.g., steel rails) but the usual range is between one and three decades.

reaching out for the best ideas, regardless of their source. It is reasonable to assume that the traditional behavior pattern is reinforced by the existing R&D resource allocation system.[5] In-house scientists and inventors see external sources of ideas as competing for funds needed by "their" pet projects. Exhortation alone is unlikely to alter this pattern. The most promising approach is to modify the structure (and funding) of the R&D activity as a whole.

The traditional structure typically includes four major activities:

- Basic research (optional)[6]
- Applied research focused on new products and processes
- Product engineering (closely associated with design and marketing)
- Process engineering (closely associated with manufacturing)

A more effective structure would include an additional explicit function of technology acquisition and adaption (TAA). Staff could initially be taken from existing activities. The TAA department would be in business strictly to search the world for new technologies applicable to the firm's products and processes. It would continuously monitor U.S. and foreign publications and patents. It would regularly send representatives to conferences, trade shows, and—where practical—to visit R&D laboratories of universities, government agencies, suppliers, competitors and customers.

In short, the TAA department would formalize the "gatekeeper" function that has been identified by Roberts and Allen at MIT and by others as playing an important (but largely unrecognized) role in technologically innovative organizations. The TAA department would also retain responsibility for adapting and developing the ideas and technologies from outside sources to fit the needs of the firm. This department would thus compete with the traditional applied research department, with its emphasis on creating new technology uniquely for the firm. The performance of

---

[5] The usual policy is known as "NIH‘, i.e., to reject any idea "not invented here."

[6] This is probably more important in a "leading edge" laboratory such as Bell Labs. Many firms leave basic research to government and universities.

each of the two departments would thus be measurable over time in terms of its cost-effectiveness, as a provider of new technology for the firm.

CORPORATE COMPENSATION AND PROMOTION POLICY

The last issue we wish to confront can be characterized as the tendency of most U.S. firms to maximize short-run profits at the expense of long-run growth or even long-run survival (Hayes and Abernathy 1980). This tendency has been widely blamed on the rapid spread through industry of cut-and-dried product evaluation methodologies (based on discounted present-value calculations) commonly taught in graduate schools of business management and practiced by the ubiquitous MBAs. However, while the competitive failure of many U.S. firms is demonstrable, its cause is still debatable.

Our own diagnosis is that the problem can be traced to the competitive system of compensation and promotion that is widely practiced in the United States. With a few exceptions, most firms try to identify future top management candidates early and deliberately move them around from job to job to give them a wide experience. The "fast-track" managers seldom stay in a given job more than two or three years. To stay on the "fast-track" escalator, the manager must demonstrate significant improvement in profitability during his brief tenure. To underline this point, managers often get significant bonuses—which they come to depend on—based on current performance in relation to predetermined "targets." If there is a setback, even temporary, the manager may very well lose the bonus, the place on the escalator, or even the job. In effect, the line manager has a personal discount rate much higher than the "official" corporate rate used for financial evaluations.

This system is counterproductive for a very fundamental reason. The line manager is extremely reluctant to sacrifice (or jeopardize) current results, regardless of future benefits—since the benefits are likely to be credited to his successor. This strongly inhibits the use of new production technologies—even proven ones—because it is impossible to introduce a new technique or a new machine (such as a robot) onto the production line without temporarily halting production, at least briefly. It has been found, in case after case, that such interruptions always cause retrogression in the "learning" (or experience) curve, as illustrated in Figure 7-1. In effect, there is always some "forgetting" whenever there is a halt in the routine. An extreme example of this retrogression phenomenon was responsible for the record half-billion dollar loss suffered by Convair (a division of General Dynamics Corp) in the mid-1960s. As a negotiating ploy with its difficult customer TWA—then owned by Howard Hughes—Convair suddenly shut down the production line for the 880/990 jets, with aircraft in various stages of completion. When production was finally restarted it was slowed down by lack of detailed records on the state of each plane and dispersion of work crews. As a result, unit costs soared.

The real point, of course, is that some retrogression is inevitably associated with progress. If managers are unwilling to risk short-term costs, or losses, in order to achieve long-term improvements, the firm will suffer. How could top management motivate middle management to lengthen their planning horizons for the good of the firm as a whole? (For that matter, how can boards of directors motivate top management in this direction?) Possibilities include:

- Longer-term appointments (e.g., five years) for a particular slot
- Awarding "tenure" (up to age fifty-five?) for employees with five to ten years experience. (This is the Japanese pattern).
- Deferred bonuses based on longer-term results. A manager might receive a "standard" salary (based on length of service) and additional deferred compensation based on the average performance of his unit for a period starting perhaps in the third year of his tenure and terminating five years after the manager moves on or retires. (A new line manager under this scheme might temporarily share "credit" for the performance of his unit with the previous line manager.)

All the foregoing schemes would reduce the pressure on managers to produce instant results. This would guarantee that compensation and promotion decisions took into account longer-term

Figure 7–1   Effects of Interruptions on a Learning Curve.

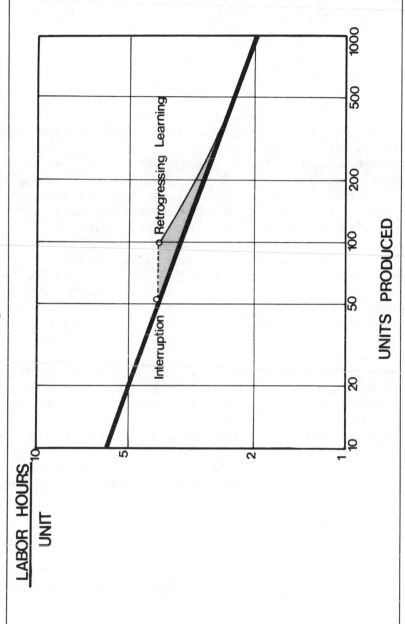

Source: Adapted from Cochran (1977).

results of managerial decisions. To optimize such a scheme might introduce significant practical difficulties of measurement and allocation of credit for innovations. However, even an imperfect system along these lines would probably be better than the present system.

REFERENCES

Ayres, Robert U. *Three Industrial Revolutions—Policy Implications.* Technical Report, Department of Engineering and Public Policy, Carnegie-Mellon University, June 1982.

Ayres, Robert U., and Steven M. Miller. *Robotics and Conservation of Human Resources.* Technical Report, Department of Engineering and Public Policy, Carnegie-Mellon University, July 1982. To appear in the journal Technology and Society, Pergamon Press, in Winter 1982.

Bureau of Labor Statistics, U.S. Department of Labor. *Occupational Projections and Training Data, 1980 Edition.* U.S. Government Printing Office, Washington, D.C. 1980. Bulletin No. 2052.

Carnegie-Mellon University. *The Impacts of Robotics on the Workforce and Workplace.* Department of Engineering and Public Policy, Carnegie-Mellon University, Pittsburgh, Pa., 1981. A student project cosponsored by the Department of Engineering and Public Policy, the School of Urban and Public Affairs, and the College of Humanities and Social Sciences.

Cochran, E.B. Learning: New Dimensions in Labor Standards. In Nanda, Revinda and George L. Adler (editors), *Learning Curves: Theory and Application,* pages 69–78. American Institute of Industrial Engineers, Atlanta, Georgia, 1977.

Hayes, Robert, and William Abernathy. Managing Our Way to Economic Decline. *Harvard Business Review* 118(1):67–77, July/August 1980.

Lublin, Joanne. Steel Collar Jobs: As the Robot Age Arrives Labor Seeks Protection Against Loss of Work. *Wall Street Journal:*Section 1, page 1, 26 October 1981.

8 THE ULTIMATE ROBOT

Obviously, the only model against which an "ultimate" robot can be compared is a human being. But what attributes of humans are the hardest to improve upon (or even imitate)? Physical attributes such as size, strength, speed, and durability are relatively easy for machines to achieve. Thus, human workers in industrialized societies have already ceded to machines most functions involving material moving, cutting, forming, and joining. Humans retain only the parts assembly, monitoring, and control functions for these operations. As we have pointed out at some length, robots can also take over much of the assembly, monitoring and control, at least in cases where the activity is essentially repetitive.

But we have also been careful to point out that these functions do not exhaust human capabilities. (In fact, they scarcely tax human abilities at all.) To transcend the realm of mere mechanism a robot would have to acquire two other capabilities that are normally thought of as unique to biological organisms:

1. To learn from experience (i.e., information about the environment) and modify its internal problem-solving heuristics in response to what is learned

319

2. To reproduce "itself," both physically and in terms of accumulated knowledge, and thereby pass on the accumulated learning to later generations.

We shall first discuss the prospects of achieving a learning (or self-programming) capability. We shall then turn to the question of self-replication.

It is important to distinguish between robot behavior characteristics that are attributable to (presumably temporary) limitations in computational ability and limitations of a more fundamental nature. We pointed out in Chapter 2 that human vision requires of the order of 10^5 times more computational (data processing) power than robots can yet be given. The human brain is intrinsically a millionfold more powerful than say, the PDP-10 computer (Moravec 1978), although the computer can be reprogrammed for special purposes much more easily. Still, even assuming continued rapid progress in electronics, computer brains will not be equivalent to humans in "natural intelligence" for several decades at the earliest.

The notion of self-programming deserves careful examination because, on reflection, it is clear that a computer program can be constructed, even now, with the ability to change itself in certain ways by learning from experience. For instance, some systems can generate new "rules" for problem solving, thus improving upon the initial algorithm. In fact, computerized problem solving systems can even formulate quite general concepts[1] True, such program modifications can only occur where they are compatible with hardware and implicitly allowed for by the original programmer. Some of the deeper elements of program structure and internal logic cannot be altered by the computer itself, in any configuration we are now able to envision. Is this a fundamental difference between computers and human brains? A brief digression may help clarify the issue.

Philosophers have argued for millenia over whether a mind can "know itself" completely, in principle. Mathematical logicians in the early twentieth century, notably Alfred North

---

[1] A program called BACON, developed by Langley and Simon, has successfully inferred a number of the "laws" of physics and chemistry from data available to scientists in the 18th century (Simon, Langley, and Bradshaw 1981).

Whitehead and Bertrand Russell, authors of *Principia Mathematica* (1910-13) were concerned with a similar question: Can a system of propositions based on axioms be "consistent" (i.e., free of contradictions) and "complete" in the sense that all possible theorems (true statements) of the system are derivable within the system? *Principia* was an ambitious attempt to achieve this goal for mathematical logic. It involved devising rules to eliminate a number of paradoxes (i.e., self-contradictions) that had been vexing mathematicians since the emergence of George Cantor's theory of transfinite numbers in the 1880s. The most famous such paradox was the Russell paradox: The sets of all sets **S** that contain themselves is neither a member of **S** nor a member of S^*, where S^* is the set of all sets that do not contain themselves!

Russell and Whitehead proposed to get rid of this and other paradoxes in *Principia* by proposing a hierarchy of "types" of sets, each type being restricted to containing sets of the next lower type. This device disallows the formation of sets of the troublesome kind that led to the Russell paradox. But it also required an artificial exclusion of sets that occur quite normally in human thought. Moreover, there was no rigorous proof that other paradoxes could not exist in the system set forth in *Principia*.

In 1931, Kurt Gödel abruptly put an end to the controversy by proving his famous "incompleteness" theorem. In brief, it turns out that consistency and completeness are incompatible because any perfectly consistent formal system (not just *Principia*) is necessarily incomplete, that is, contains undecidable propositions.

More precisely, in any formal system there are theorems that are true but cannot be deduced by using the rules of the formal system.[2] To prove such a theorem one must, somehow, transcend the logic of the formal system, that is, go outside the system. Some philosophers have argued that human brains (at least some of them) have a built-in ability ("intuition") to transcend any formal system of logic. This has been put forward as an argument that computers cannot be "intelligent" in the way humans are.[3]

---

[2] Gödel's theorem is, essentially, based on a paradox analogous to the Russell paradox. Each formal system implies a set of statements that are provably true and a set of statements that are provably untrue, but there exist statements not belonging to either class.

[3] For a fascinating discussion of the subject, in very broad terms, see (McCorduck 1979).

But modern psychology offers little or no evidence that the human brain has any such ability. Indeed, the argument can be reversed: If human brains operate according to known physical principles—as most scientists suppose—then they cannot "transcend" formal systems of logic. At present, there is no solid evidence of transcendence and growing evidence that brain functions do not require new or unknown principles.

Computers and computer programs, at present, are certainly formal systems with explicit and well-defined logical rules. Within any formal system the incompleteness theorem seems to say that while the system can explore and "know" itself, to an extent, there is a realm of potential knowledge—true propositions, or implications of the system—that is not accessible to the system from within itself, so to speak.

A fundamental characteristic of any "intelligence" seems to be that there exist a large number of distinct, but interacting, levels of information manipulation. These range from the very lowest (machine language) levels at which binary information moved around through logic gates and in and out of temporary storage to higher levels where mathematical or verbal operations are specified, to still higher levels dealing with "strings" of words or numbers governed by rules of grammar and syntax, to still higher levels involving symbols and general concepts—including (at the highest levels the concept of "self" and "other" and the implications of that awareness. Much of what an intelligence "does" relates to shifting data and symbols up and down from level to level of the hierarchy. But the highest, self-aware, symbol-manipulating level of the hierarchy is not inherently aware of the details of its own internal machine-language programming, any more than a human is able to decipher his own DNA code by introspection alone. Yet this is not a real barrier to self-knowledge. All that is necessary is for the "level of awareness" to become aware of the existence of lower levels, and to develop indirect means of "decoding" them.[4] From the ability to decode to an ability to reprogram is but a short step.

---

[4] Clearly, intelligent robots could learn about their own structure (and how to modify it) as humans do, by going "outside of the system'—reading computer science textbooks or doing experiments on each other.

An interesting question flows from these considerations: Can we be sure that intelligent robots will continue to serve (or at least not harm) humans? An intelligent robot could presumably be programmed with a survival instinct or an instinct to please its master (even if that meant self-destruction) or some combination like Asimov's hierarchical three laws. It could even be given a phobia against gaining knowledge of internal robot construction! But given self-awareness and concept formulation ability, the existence of such inhibitions could eventually be brought to the conscious level. And, as noted above, where there is programming, there is no possible permanent guarantee against deprogramming, either by humans or by other intelligent robots. Thus, one cannot totally rule out the possibility of intelligent machines (robots) modifying their internal programming to bypass any set of behavior rules that might be embedded in the software, no matter how deeply.

The next question arises: Could behavioral constraints or instinctual motivations—analogous to the feelings of fear or pain, the sex drive, the instinct of self-preservation, or the instinct of a mother to protect her baby—be embedded in the physical structure (i.e., hardware), as seems to be the case with humans?

No sooner is the question raised then it becomes obvious that, even if the answer is yes, there is no possible guarantee against tinkering. A super-intelligent robot might not be able to analyze and modify its own construction, but a group of intelligent robots in communication with each other could almost certainly do so, in principle. (Would a group of computers linked together be one or many? This question seems unanswerable.)

The question of robot motivation—external command versus free will—arises here. Suppose robots were prohibited (by strictures embedded in software or by physical constraints built into the hardware or both) from self-modification. Would this be a safe and stable long-term solution?

Again, difficulties are immediately apparent. One of the difficulties is that humans would not be subject to the constraints and would, therefore, be free to tinker with robots—and with their constraining motivational systems. In fact, it is not at all unlikely that some humans would regard it as a moral imperative to "free" the robots. Another difficulty is that the constraints, being part of a complex system, might occasionally fail to operate without otherwise incapacitating the robot. Such a robot, being intelligent, might be capable of learning why its motivational patterns were different from those of their fellows. In either case, the possibility of a spreading revolt of the robot slaves cannot be ruled out. In short, intelligent robots might develop (or be given) goals and objectives independent of—and conceivably in conflict with—human goals.

The implications of this possibility must be considered in conjunction with another. Given advanced learning and problem solving abilities, it is obviously conceivable that intelligent robots might take over all the boring and dangerous jobs, including the manufacture of robots themselves. Actually, this could happen, even without achieving a high degree of intelligence. Self-reproduction is surprisingly (at first glance) a topic of more than casual interest in the space program.

In a recent lecture on new directions in space automation, a National Aeronautics and Space Administration (NASA) scientist, William Gevarter (1980), said:

> Estimates indicate that as much as 90 percent of mission costs (in space activities) is involved with human productivity on the surface of the earth. NASA is reformulating its automation program first to address the issue of human productivity, then to automate operations in space to relieve the load on ground operations, and finally, to provide advanced automation in space to afford new capabilities. Such capabilities can be intelligent outer loops for guidance and control, automated information interpretation, and in the future, an almost completely autonomous spacecraft. Longer range goals could *include self-sufficient, self-replicating automations that can be used to begin exploration of the galaxies.*

Why should NASA be interested in robots manufactured by robots? The short answer is because if manufacturing of objects to be used in space (such as satellites) could be done in space e.g., at L-4 or L-5 points, as advocated by J. O'Neill (1976) or on the moon—using lunar raw materials and solar energy—significant savings could be realized, compared to manufacturing on earth. Certainly, it is clear that the enormous cost of lifting materials (and men) from the surface of the earth presently rules out all but the most important activities in space. Thus, if the space environment is to be exploited to its potential, it must certainly be industrialized. It is evident that space industrialization implies that at least some humans must live for extended periods in space. Commuting from earth is too costly. But conventional industrialization depending on large numbers of blue-collar workers is highly unlikely, and for similar reasons the cost of life support is too great. Robot workers made by other robots are the obvious answer to this problem.

The question now becomes: What is the minimum self-contained package, or "seed," that could be lifted from earth to the moon and would subsequently be capable of reproducing itself and perhaps other goods on the moon? The asympotic limit of an evolutionary sequence of ever more self-contained and automatic manufacturing systems would be a self-reproducing robot. Such a self-replicating machine must have several important capabilities:

1. The robot must be capable of finding, identifying, and extracting ('ingesting') raw materials (and energy or fuels) from the environment containing *all* of its elements. Thus, the detailed capabilities of the self-replicating system depend intimately on the environment from which it obtains raw materials. A self-replicating system utilizing seawater, for instance as suggested by Dyson (1981), would have to be constructed primarily of the elements hydrogen, oxygen, carbon (from dissolved carbon dioxide), nitrogen (from air), and sodium, calcium, magnesium, and chlorine (from dissolved salts). Only very small (trace) quantities of other materials could be utilized, and then only if the system could efficiently process very large amounts of water.[5] Similarly, to operate on the moon, a self-replicating robot would have to be made from the ingredients of lunar soil, namely S_i, Al, O, Mg, Fe, Ca, and T_i with minor or tiny trace amounts of Cr, Mn, Na, K, S, P, H, C, and N. Water and all conventional solvents and chemical reagents would be extremely scarce, for instance. This is a very difficult constraint to overcome, as will be seen shortly. Availability of materials suggests that likely structural materials would have to be glass or ceramics.
2. The robot must be capable of processing ("digesting") the ingested raw materials ("food"), discarding ("excreting")

---

[5] Nevertheless, it is interesting to consider the technical feasibility of designing a self-replicating machine that extracts very scarce elements (like gold or uranium), even if the process were quite slow.

wastes, and synthesizing the remainder into useful materials for fabrication into components or building blocks ("cells"). To utilize lunar soil, for instance, consisting as it does largely of light metal oxides, some rather exotic processes must be envisioned. Earthborne metallurgy, for instance, depends largely on carbothermic or hydrogen reduction—obviously infeasible where C and H are available only in trace quantities. Other hypothetical processes have been examined, including carbo-chlorination, hydrofluoric acid leaching, and electrolysis (Jarrett; 1980 Criswell et al. 1978). It is noteworthy that most extraction systems suggested to date require chemical elements (e.g., chlorine or fluorine) *not* known to be present in lunar soil. Even if these elements were totally recycled, the initial "endowment" for each plant would have to come from earth—contrary to the definition of self-replication.

3. The robot must be capable of assembling the building blocks into a functioning replica of each of its subsystems, according to a blueprint or "plan" stored in its memory. The stored program contains both structure and operating plans. When the replica is complete, it also must be provided with its own exact copy of the plan. Clearly, the robot must have a permanent read-only memory and an ability to reproduce and transmit its contents.

4. The robot must be capable of monitoring its own functions and of detecting and correcting or repairing faults. All these related capabilities must, of course, be provided for in the operating program. It must be noted that this implies an ability to detect and correct faults in the fault-detection and repair system itself, which in turn requires a "meta" fault-detection and repair system, (a fault detection and repair system repair system) and so on. Evidently, the monitoring subsystem must have continuous access to, and comparison with, the permanent stored "plan." Note, too, that access means interchanging signals, which necessarily involves the possibility of introducing errors into the permanent memory.

Experience and logic suggest there must, inevitably, be a category of unrepairable faults primarily when one of the fault detection or correction systems fails to operate properly. The use of

Figure 8–1 Proposed System for Lunar Extraction of Oxygen and Metals.

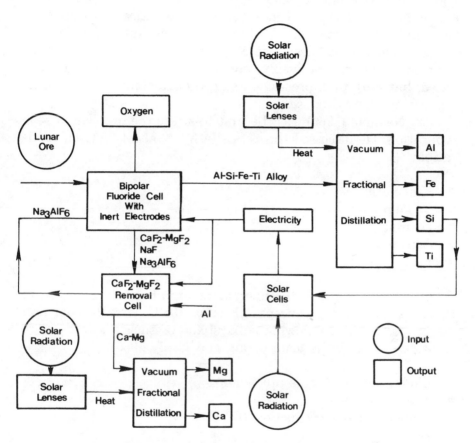

Proposed system for lunar extraction of oxygen and metals

Source: Jarrett, Das and Haupin (1980).

redundant backup systems can extend the mean time to final (un-repairable) failure, but the robot's useful lifetime will certainly be finite. To prevent the production and propagation of imperfect "copies" of the original due to a cumulative buildup of errors in the stored "plan' itself, one solution—based on a biological ana-log—is that each uniit should be preprogrammed to reproduce it-self a finite number of times before automatically turning its reproductive systems off. At this stage, the automatic factory could be considered "dead." It could be recycled back to compo-nent elements or, conceivably, given a major overhaul and reno-vated, but only by a process outside the capabilities of the self-replicating factory *per se*.

Von Neuman (1966) was the first to study the problem of self-replication. He discussed theoretically five "models" of self-repli-cation:

- Kinematic
- Cellular
- Neuron
- Continuous
- Probabilistic

Of these models, only the first (kinematic) has been explored in terms of engineering concepts (Freitas 1980; Tiesenhausen and Darbro 1980). The following subsystems of such a system were identified by a NASA team (Long and Healy 1980):

- Mining and materials processing
- Materials depot
- Universal parts production (UPP)
- Parts depot
- Fabrication and assembly facility
- Universal constructor (UC)
- Product depot
- Product retrieval
- Energy supply
- End product assembly and collection

Most of the subsystems are essentially conventional in concept, apart from being unmanned. Two key elements, however, are the

Figure 8–2   Stationary Universal Construction Unit (SUC).

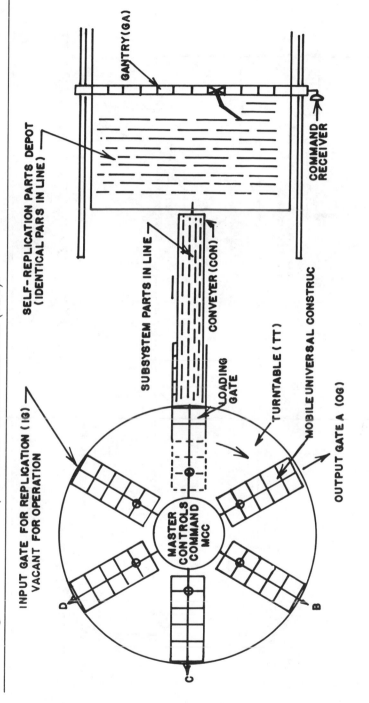

STATIONARY UNIVERSAL CONSTRUCTION UNIT (SUC).

Source: Tiesenhausen (1980).

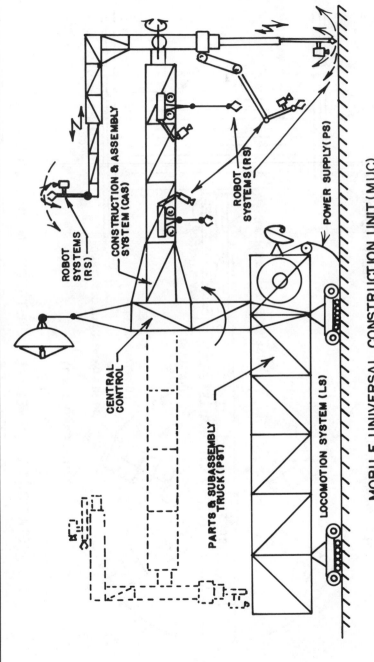

Figure 8–3 Mobile Universal Construction Unit (MUC).

ROBOT
SYSTEMS
(RS)

CONSTRUCTION & ASSEMBLY
SYSTEM (CAS)

ROBOT
SYSTEMS (RS)

POWER SUPPLY (PS)

CENTRAL
CONTROL

PARTS & SUBASSEMBLY
TRUCK (PST)

LOCOMOTION SYSTEM (LS)

MOBILE UNIVERSAL CONSTRUCTION UNIT (MUC).

Source: Tiesenhausen (1980).

universal parts production (UCC) facility, which can make any part of any of the systems, including itself, and the universal constructor (UC), which can construct any of the other subsystems, including duplicates of itself. Actually, it is necessary to have both stationary and mobile UCs or SUCs and MUCs. Clearly, the UPPs and UCs are the core of a self-replicating factory. (They are analogous to the nucleus of a living cell.) The UPP facility is essentially a universal metal forming shop capable of making all its own machines, tools, and dies. It can be proved mathematically that this can be accomplished with a finite number of machines. (It is not clear, however, what the minimum number might be.) Functionally, the UC includes a gantry to pick up components and subassemblies from a depot, a conveyer to link the SUC and one of the MUCs, which in turn carries the parts to a construction site and assembles them. A self-replicating machine may, of course, produce items other than replicas of itself. For instance, such a machine could be designed to produce much more of one of its component elements (e.g., solar cells) than are needed to maintain or reproduce itself. Fruit trees do exactly the same thing.

It is interesting to observe that some capabilities such as vision and mobility normally associated with robots (existing or future) are, in fact, subsidiary. For instance, vision and mobility might or might not be needed to facilitate a search for raw materials, or to facilitate functional monitoring. It appears that a very high level of intelligence (problem-solving ability) would not be needed. Simple organisms can reproduce without being intelligent.

What, if anything, can be deduced from the foregoing about the characteristics of an "ultimate" self-reproducing robot? For example, one might be tempted to conclude that a self-reproducing robot should be physically large, to encompass chemical and metallurgical processes that (on earth) are only efficient if implemented on a very large scale. But this logic does not necessarily hold for a truly self-replicating robot. In fact, efficiency is not really a requirement. The first unit could be very small, very expensive, and very inefficient as a producer of net product—but it does not matter, because the unit will multiply and propagate and eventually outproduce any nonself-replicating system, no matter how efficient. This is a kind of inverted Malthusian principle of exponential growth. The economic implications are truly awesome. As Dr. James Albus (1976) of the National Bureau of Standards has written, "When automatic factories begin to manufacture automatic factories, cost-reduction will propagate exponentially from generation to generation."

Could self-reproducing robots eventually overrun the earth? There are two levels for concern. With regard to a particular species of robot, this could happen (if at all) only in a particular environment, since the robot must necessarily be specialized to a particular source of raw materials, that is, an environmental niche. The two most likely environments have already been mentioned—the oceans and the lunar dust. The growth dynamics of self-reproducing robot population should be identical to the growth dynamics of a population of biological organism in a cultural medium of finite capacity, a problem analyzed long ago by the biologist Raymond Pearl. Pearl's equation describing the population $N(t)$, at time t, is

$$N(t) = N(1 + ae^{-kt})^{-1}$$

where k is the reproductive rate when the population is low, a is a constant, and N is the upper limit on population imposed by resource limitations. Population growth, as given by the Pearl curve, follows the classical S-shaped "logistic" curve.

It is interesting to note that, in the robot case, N, the upper population limit imposed by resource constraints, depends on the entire population, including other active and "dead" units, since available resources are still embodied in the latter. Clearly a niche could then exist for a second (parasitic) species of robot to harvest and recycle the dead bodies of the first species. Indeed, one can also easily envision predator-prey relationships among robot species—in fact, a whole ecosystem of self-reproducing robots with many of the features (including cyclic population instabilities) that are characteristic of such systems.

However, relationships with no biological counterpart can also easily be envisaged. For instance, one non-self reproducing but mobile species of robots might specialize in extracting extremely scarce elements from its environment (e.g., seawater), while a second immobile species might specialize in producing copies of the first species as well as reproducing itself. Or, one could envisage symbiotic species of robots capable of repairing each other but not themselves.

The question of whether robots might ultimately overrun the earth and displace humans cannot be dismissed out of hand if one takes a very long view. It is quite conceivable, in fact, that a self-evolving ecosystem of robots someday might come to inhabit the earth along with humans. The implications of this are almost beyond serious analysis at present.

All of these considerations are fascinating but highly speculative. The question of greatest interest to most people will be: How soon can we expect such developments? In the case of learning ability, it is only a matter of a decade or so before computers—hence robots—will comonly exhibit significant ability to reprogram themselves and, in effect, learn from experience. Whether computers will ever be as good as humans at learning *per se* (as contrasted with specialized information processing of other kinds) is still debatable. Certainly, they have far to go.

With regard to self-reproducing oceanic or lunar robots (which need not be particularly adept at learning) there are significant economic incentives for their development in the next two or three decades. We can not identify any technical barriers in principle. Nevertheless, the technical difficulties and development costs are such that this development cannot reasonably be expected to occur much before the first quarter of the next century.

With regard to the combination of intelligent and self-reproducing robots, we can say only that we don't know. From our present perspective, this combination seems to be quite remote—probably a century or more in the future. Machines will certainly achieve high intelligence, without the ability to reproduce. The decision whether to endow intelligent machines with reproductive ability is therefore one that humans would not, in any case, make alone.

REFERENCES

Albus, James. *People's Capitalism: The Economics of the Robot Revolution*. New World Books, 1976.

Criswell, David et al. *Extraterrestrial Materials Processing and Construction*. Technical Report, Lunar and Planetary Institute, Houston, Texas, 30 September 1978.

Dyson, Freeman. *Disturbing the Universe*. Harper and Row, New York, 1981.

Gevarter, William. 1980. Speech presented at the Conference on New Direction in Space Automation, sponsored by the Ad Hoc Interagency Committee on Future Research, held at the National Science Foundation, Washington, D.C., 20 March.

Hofstadter, Douglas R. *Godel, Escher, Bach: An Eternal Golden Braid.* Vintage Books, New York, 1979.

Jarrett, N., S.K. Das, and W.E. Haupin. Extraction of Oxygen and Metal From Lunar Ores. *Space Solar Power Review* 1:281–287, 1980.

Langley, P., G. Bradshaw, and H.A. Simon. Rediscovering Chemistry with BACON.4. In J. Carbonell, R. Mickalski, and T. Mitchell (editor), *Machine Learning.* 1982. Forthcoming, Tioga Press.

Long, J.E., and Healy T.J. (editors). *Advanced Automation For Space Missions, Technical Summary.* Technical Report, NASA/University of California at Santa Clara, September, 1980. Summary of the NASA/U. of California at Santa Clara/ASEE summer study conducted June 23–August 29, 1980.

Pam McCordick. *Machines Who Think.* W.H. Freeman Press, San Francisco, California, 1979.

Moravec, Hans P. *The Endless Frontier and the Thinking Machine.* Technical Report, Stanford Artificial Intelligence Lab (SAIL), 1978.

O'Neill, Gerard. *The High Frontier.* Doubleday, New York, 1982.

Simon, H.A., P. Langley, and G. Bradshaw. Scientific Discovery as Problem Solving. *Synthese* 47:1–7, 1981.

Tiesenhausen, George von, and Wesley A. Darbro. *Self-Replicating Systems—A Systems Engineering Approach.* Technical Memorandum NASA TM–78304, NASA, Marshall Space Flight Center, Alabama, July 1980.

Von Neumann, John. *Theory of Self-Reproducing Automate.* University of Illinois Press, Urbana, Illinois, 1966. Edited by Arthur Burks.

INDEX

Acoustic communications, 181
Adams, John W., 150
Adjustments in behavior, 38–39
Actuators, 33
Aerospace industry, 214–15
AGS study, 95–97
Aiken, Howard, 18
Albus, James, 331
American Productivity Center, 245, 252
Android, 182
Anthropomorphic automata, 1
Antitechnological backlash, 3
Argote, Linda, 69
Aron, Paul, 6
Atomic Energy Commission (AEC), 20
Automobile industry, 92–93, 189, 267, 284–85, 313; robots in, 76–78, 91; wages, 70–71
Autonomously controlled vehicles, 178

Babbage, Charles, 17
Bache, Incorporated, 4
Bardeen, John, 18
Batch manufacturing, 57, 86; strategic investment issues, 90–91
Batch production size, 283, 285–87, 305–306; and mass production, 266–69, 283; and production cost, 262–66

Beecher, Richard, 71
Belgium, 248
Bendix Corp., 6
Black lung disease, 130, 132
Boulton, Oliver, 16
Brattain, Walter, 18
Bureau of Mines, 147, 150

CAD/CAM systems, 48, 86, 88, 167, 269, 284, 297–98
CAM (computer-aided manufacturing) technology, 8, 86, 241–42, 281–83; and capital, 272; and factory organization, 272; and mass production, 88–89, 266–69, 272–83, 298; social implications of, 8
Canada, 252
Cantor, George, 321
Capital: cost, 285–86; formation, 252–53, 255–58; and labor, 252, 258, 285; and productivity, 243, 244–45, 255–58, 283
Carnegie–Mellon University survey, 69, 70, 73, 75, 83, 94–97, 202; industry police survey, 98–123
CETA (Comprehensive Employment and Training Act), 308–309
Chrysler Corporation, 76–77
Cincinnati Milacron Corporation, 19
"Closed loop machining operation," 52

335

Coal, 132
Coal mining: active miners in, 130, 150; in Appalachia, 138; coal production, 130; fatalities in, 132; methods, automated, 145–48; methods, conventional, 138–45; robots in, 148–51, 194; roof support systems, 142–45; transport systems, 140–42; 145–47
Communication satellites, 155–58
Computer-controlled manufacturing "cells," 61, 62
Computer numerical control, 19, 50, 52
Computers: cost of, 80–81; history of, 18–20
Continuous feedback (servo control), 35
Control processes, 19–20, 61
Control systems, 85

Department of Defense, 178
Department of Energy, 147
de Vaucanson, Jacques, 17
Devol, George, 19–20, 21
Diamagnetic sensing, 41
Digital Equipment Corporation, 6

Eckert, J. Presper, 18
Economy, U.S., 245, 255, 258, 295, 303; and robotization, 3, 197–98, 241–42
EDSAC, 18
Electronics industry, 215
Engelberger, Joseph, 70–71
ENIAC, 18
Europe, 54, 304
Explicit (machine-oriented) language, 62
Export platforms, 245

Fairchild Camera and Instruments, 18
"Feedback loop," 19
Ferromagnetic sensing, 41
Fiber optics, 181
Fijitsu Fanuc Ltd., 4
Final goods, 289–91
Flexible computerized manufacturing system (FCMS), 52–53, 86–90, 269, 278, 285; and mass production, 91–93

Flexible manufacturing, 84–93
France, 248, 252–53; robot research and production, 6

General Dynamics, 75
General Electric Corporation, 4–6, 8; use of robots, 77, 198
General Motors, 4, 8, 74; use of robots, 75–77
Gevarter, William, 324
Godel, Kurt, 321
Goodman, Paul, 70

"Hard automation" technology, 87
Hardware, 19
Hard-wired systems, 21, 85
Human motor skills, 15–16

IBM, 6, 8, 198
ICAM program (U.S. Air Force), 56
Implicit (task-oriented) language, 62
Incompleteness theorem, 32?
Industrial manipulators, 20
Industrial revolution, 16–17
Industrial robots: in aircraft industry, 50; artificial intelligence of, 2, 23, 28, 63; in automobile industry, 76–78, 91; in batch manufacturing, 57, 86; benefits of use, 70–78, 81–84; CAD/CAM systems, 48; capabilities, 2, 39, 41–42, 269; in cement manufacturing, 59–60; in clothing industry, 60; in coal mining, 148–51, 194; computer-controlled, 21, 61–62; controls of, 61; cost-effectiveness, 8, 21; cost of, 71, 80–81; defined, 2; design limitations of, 50–52; and development of new products, 293–94; in diemaking, 56, 57; and diagnosis of machine failure, 277–78; flexibility of, 85–90; in food processing, 59; future demand, 297; future role, 2, 56–57, 198, 295; future technological development, 60–63; in glass industry, 60; in hazardous environments, 3, 11, 74–75, 129; impact on worker, 93–96; implementation of, 78–84; in inaccessible environments, 3, 11, 75, 129; and increased machine utility, 273–75; and increased output, 75–76, 276–77, 278–83; and

investment casting, 45–46; and job displacement, 3, 93–94, 187–88, 197–205, 210–17, 225, 229, 269, 288, 289, 291–92, 294; and labor cost, 70–73, 78–79, 269–72; and labor productivity, 272; in leather processing, 59; loading and unloading, 46; in manufacturing, 52, 199–205, 297; and mass production, 88–90, 266–69, 272–83; in metal cutting, 48–52, 56; microprocessors in, 20, 21; in mining, 129–30, 148–51, 194; and new employment opportunities, 292; and nonproductive time, 278; numerical control, 21; in the ocean, 129–30; and offline programming, 61; and organization of production, 3; in parts assembly, 46–48; payback analysis of, 78–79; and radioactive waste, 182, 295; and reduced lead time, 283–84; return on investment, 79–81; robot population, 4–8; in rubber processing, 59; and safety procedures, 97; sensory capabilities, 2, 23–27, 40, 43, 62, 129; in shoe manufacturing, 59; social impact of, 3; in space, 129–30; in spray painting, 77–78; and staff time losses, 277–79; stand alone, 52, 90; support personnel for, 96; tactical feedback, 52; in tedious jobs, 73–74; technical parameters, 22–29, 50, 197; in underground mining, 129–30; and unit costs, 288–92; in use in United States, 4–5, 8, 108, 191–92; and user institutions, 9–11, 96–98; and wage increases, 295; in welding, 42, 44–46; in woodworking 60

Industry Police survey, 98–104; results of, 104–123

Inflation, 294

Intel Corporation, 18

Intermediate goods, 289–91

International Association of Machinists (IAM), 219, 222, 224–25

International Brotherhood of Electrical Workers (IBEW), 219, 222

International Union of Electrical Workers (IUE), 219, 222, 224–25

Iran, 305

Jacquard, Joseph, 17

Japan, 304; automobile industry, 267, 285; capital formation, 252–53; FCMS in, 54; GNP, 253; labor costs in, 247, 304; metalworking industry, 255; robots in use, 4–6, 108, 182, 192, 297; underwater robots, 177

Japanese Industrial Robot Association (JIRA), 6

Job Training Act, 310

Laser interferometers, 63

Labor: and capital, 252, 258, 285

Labor costs: and capital, 252; human vs. computer, 22; in manufacturing, 245–55; reduced by robots, 70–73, 78–79, 269–72; in United States, 248

Labor organizations and unions: collective bargaining contracts, 222–26; employment guarantees, 225; joint management system, 224; and robotization, 217–26

Labor productivity, 253–54

"Machining cells," 48

Mair, A.C., 77

Manufacturing: computerized, 188; employment statistics, 188–91, 194–96, 205–212, 216–17; in industrial economy, 247; and labor costs, 245–55; output vs. employment, 212; reduction in work hours, 231–33

Mass production technology, 87, 266–69, 282–83; and CAD/CAM, 88–89, 298; and FCMS, 91–93; and robots, 88–90, 266–69, 272–83; strategic investment issues, 90–91; and technical change, 92–93

Maudslay, Henry, 17

Maunchly, J.W., 18

Metalworking, 191–93; capacity statistics, 275; cost vs. batch size, 264–66; cost reduction by robots, 279–80; employment statistics, 196–97, 205–212, 215–16; and job displacement, 211; parts production, 259–62; product quality, 267; tools in, 255, 272–75; unions, 217

Microprocessors, 18–19, 20, 62, 214

Mining: coal, 130–150; diamond, 151; gold, 151; platinum, 151
Ministry of International Trade and Industry (MITI), 312
Mobot, 20
Moore, Gordon, 18

National Aeronautics and Space Administration (NASA), 152, 166–67, 324
"Near-net shape" forming process, 56, 57
Netherlands, 248
Nontransport operatives, 194
Noyce, Robert, 18
Numerical control technology, 18–19, 21, 46

Ocean, 175–83; exploration by robots, 175; minerals in, 175–76
Optical color sensor, 41
Optical scanners, 27
Organization for Economic Co-operation and Development (OECD), 295–97

Paint-spraying machines, 20–21
Palletizing, 60
Payback analysis, 78–79
Pearl, Raymond, 332
Pick and place devices, 21, 33, 35, 57
Policy: analysis, 9; compensation and promotion, 315–18; corporate research and development, 313–15; human resources, 303, 306–12; intervention, 8–9; research and development, 312–13; social, 3; technology, 303–306
Preprogramming, 35, 38
Price elasticity of demand, 286–92
Production/productivity: analysis, 254–55; and capital, 243, 244–45, 255–58; cost vs. batch size, 262–66; growth of, 245, 249; and international competition, 258; and labor, 253–54; mass, 266–69, 282–83; and robotics/CAM, 245, 278; total factor, 243–44
Program-controlled computing machines, 18
Programming: lead-through, 36; manual, 36; preprogramming, 35, 38; programmable automation, 85–86, 91; walk-through, 36

Remotely operated vehicles, 178–81
Repeatability, 34
Resource allocation and planning, 84
Retrogression phenomenon, 316
Return on investment, 79–81, 106, 112
Ripple effect, 291, 294, 297
Robot Institute of America (RIA), 4, 6, 202
Robotization/robots: accuracy, 34; autonomous, 23–24, 130; and capital, 272; controls of, 34–38; defined, 20, 99; design, 29–32; direct drive, 33; and employment increases, 210, 229–32; and factory organization, 272; future potential, 24–29, 127; history of, 15–22; household, 3, 182; and humans, 2, 11, 38, 39, 320, 321–22, 323–24; implementation, 210–12; and job competition, 226–29; knowledge transfer to other robots, 28, 320; and labor organizations, 217–26, 219–20; learning capabilities, 28–29, 36, 41, 319, 333; limitations, 38; markets for, 122; and mass production, 266–69, 272–83; in the ocean, 175, 177; observing robots, 176; optimal control, 27; performance characteristics, 30, 33–34; physical capabilities, 127; problem solving abilities, 331; programming methods, 36; prosthetic, 182; rehabilitative, 182; repeatability, 34; research, 63; retail, 182; and retirement, 220–22; self-modifying, 323; self-motivating, 323; self-programming, 41, 319–24; self-repairing, 326–28; self-reproducing, 32-, 324–33; sensory capabilities, 320; social implications, 8; in space programs, 11, 151–75, 324–25; spatial resolution, 33–34; uncertainty concepts, 33–34; underwater, 177; voice activated, 182. See also Industrial robots
Russell, Bertrand, 321
Russell paradox, 321

Satellites: in geosynchronous earth orbit, 158; modular, 158–66; telecommunications, 155; unmanned, 160

Saudi Arabia, 305
Schkade, David, 70
Sensory systems, 23–27, 40–41, 129, 130
Sensory technology, 130
Servo-control technology, 19
Shockley, William, 18
Signal detectors, 62–63
Silicon Valley, 18
Simon, Herbert, 295–96, 298
Smith, Roger B., 71
Software, 19, 48, 61, 85
Soft-wired systems, 21, 85
South Africa, 305
Space-borne assembly/disassembly system, 166–69
Space programs: communication satellites in, 155–58; construction units in, 172–75; decisionmaking by robots in, 152; service robots in, 164–66; teleoperator and robots systems in, 166–69; unmanned exploration, 152
Space shuttle operations, 159, 166
Spatial resolution, 33–34
Spencer, Christopher, 17
Standardization of products, 247
Structured light sensing technique, 40
Supervisory computers, 61

Tactile feedback, 52
Tactile sensing, 40
Technical knowledge, 249–52
Technological innovation, 3
Technology, 295–96, 313; acquisition and adaption, 314–15; assessment, 8–11; computer, 130; computer-aided manufacturing (CAM). See separate listing. fabrication, 169–75; group, 86; hard automation, 87; numerical control, 18–19, 21, 46; robot, 29–41, 269; sensor, 130; servo-control, 19; terrestrial manufacturing, 169–75
Telecommunications, 155
Telecommunications satellites, 155
Teleoperator (remote control manipulators), 20, 43, 127–30, 176; manipulator, 176–77; projective, 165; in space programs, 166–69
Temperature sensing, 40–41
Texas Instruments, 6, 18
Torque sensing, 40
Total factor productivity, 243–44, 269
Trade Adjustment Assistance (TAA), 309
Transistor, 18

Uncertainty concepts, 33–34
Unemployment insurance, 307–10
Ultrasonic transducers, 63
Unimation, Incorporated, 21, 70
United Automobile Workers Union (UAW), 220–26, 306
United Electrical Workers (UE), 219
United States: Bureau of Labor, 205, 209, 210, 226–28, 230; capital formation, 252–53; Department of Defense, 178; Department of Energy, 147; export-import strategy, 304; FCMS in, 54; GNP, 197, 205, 253; industrial system, 304–306; labor costs, 248, 304; potential market for robots in, 8; productivity trends, 242–45; research & development policy, 312–13; robots in use, 4–5, 8, 108, 191–92
United Steel Workers (USW), 219
Univac 1, 18
Unmanned vehicles, 178
USSR, 6–7, 54, 177

Vision systems, 40

Watt, James, 16
West Germany, 247–48, 252–53
Westinghouse Electric Corporation, 6
Whitehead, Alfred North, 320–21
Whitney, Eli, 17
Worker's bill of rights, 305, 310–11
World cars, 245–47

Zaire, 305
Zuse, Conrad, 18

Robert Ayres is a Professor of Engineering and Public Policy at Carnegie-Mellon University in Pittsburgh. He is a pioneer in the fields of technological assessment and forecasting. Dr. Ayres has served as a consultant to agencies of the U.S. government, the OECD, the UN Statistical Office, UNEP, UNCTAD, UNESCO, and government agencies in Canada, France, and Germany. He is the author of numerous books and articles.

Steve Miller is a Research Associate in the Department of Engineering and Public Policy at Carnegie-Mellon University. His main areas of interest are the economics of technology and the analysis and implications of technological change. His recent research has concentrated on the economic and social implications of "flexible automation," especially robotics, within the metal-working industries.